PLANT-THINKING

Plant-Thinking

A Philosophy of Vegetal Life

MICHAEL MARDER

WITH A FOREWORD BY *Gianni Vattimo and Santiago Zabala*

Columbia University Press New York

Columbia University Press
Publishers Since 1893
New York Chichester, West Sussex
cup.columbia.edu

Library of Congress Cataloging-in-Publication Data
Marder, Michael, 1980–
Plant-thinking: a philosophy of vegetal life / Michael Marder.
p. cm.
Includes bibliographical references and index.
ISBN 978-0-231-16124-4 (cloth: alk. paper)—
ISBN 978-0-231-16125-1 (pbk.: alk. paper)—
ISBN 978-0-231-53325-6 (e-book)
1. Plants—Philosophy. 2. Ontology. 3. Human-plant relationships. I. Title.
QK46.M37 2013
580—dc23
2012023675

Columbia University Press books are printed on permanent and durable acid-free paper.
This book is printed on paper with recycled content.
Printed in the United States of America
c 10 9 8 7 6 5 4 3 2 1
p 10 9 8 7 6 5 4

À memória da minha avó, Maria Inácia

Καὶ γὰρ ζῷα καὶ φυτὰ καὶ λὸγου καὶ ψυχῆς καὶ ζωῆς μεταλαμβὰνει.

—Plotinus, *Enneads* 3.2.7

Contents

Foreword

GIANNI VATTIMO AND SANTIAGO ZABALA

On August 29, 2009, the democratically elected president of Bolivia, Evo Morales, was declared the "World Hero of Mother Earth" by the General Assembly of the United Nations in recognition of his political initiatives against the destruction of the environment caused by the global hegemonic economic system. According to the president of the Assembly, Rev. Miguel D'Escoto Brockmann, the Bolivian politician has become "the maximum exponent and paradigm of love for Mother Earth."[1] However, Morales is not the only South American leader to promote the environment; together with other socialist politicians, such as Castro and Chávez, he has been consistently calling for an end to capitalism's violent imposition on the environment and for an adoption of sustainable social policies respectful of our most vital resources. The fact that Western democracies constantly delegitimize[2] such policies is

1. Rev. Miguel D'Escoto Brockmann, "Morales Named 'World Hero of Mother Earth' by UN General Assembly," *Latin American Herald Tribune* (http://laht.com/article.asp?ArticleId=342574&CategoryId=14919).
2. Examples of such delegitimization have been coming from prominent newspapers and journals such as the *New York Times*, *El País*, the *Washington Post*, and *Foreign Policy*. A detailed account of this distorted information can be found in the fourth chapter of our *Hermeneutic Communism* (New York: Columbia University Press, 2011).

an indication of their indifference toward the environmental calamities (and financial crises) their own logic of profit produces, not to mention that this delegitimization shows, most of all, how thoroughly these democracies are framed within metaphysics. If, as it seems, there is a correspondence between the liberal capitalist states and metaphysical philosophical impositions that consider the environment as something that must be manipulated for our own purposes, philosophy has the obligation to break this deplorable alliance.

Although Michael Marder does not mention these politicians in his study, the political essence of their ecological initiatives is not foreign to his philosophical endeavor, a project entrenched in "weak thought,"[3] that is to say, in the philosophy of the weak who are determined to cut the tie between politics and metaphysics. But what exactly is weak thought?

Weak thought, contrary to other philosophical positions, such as phenomenology or critical theory, has not developed into an organized system, because of all the violent consequences such systematization always entails. The violence of systems is often expressed through metaphysical impositions, which aim to submit everything to their own measures, standards, and agendas. But as the philosophy of the "weak" (which claims the right of the oppressed to interpret, vote, and live), weak thought not only follows a logic of resistance, but also promotes a progressive weakening of the strong structures of metaphysics. Weakening, like deconstruction, does not search for correct solutions wherein thought may finally come to rest, but rather seeks ontological emancipation from truth and other concepts that frame and restrict the possibilities of new philosophical, scientific, or religious revolutions. These revolutions, as Thomas Kuhn explained, are the indications that science shifts through different phases and—instead of making "progress

3. Weak thought—*pensiero debole*—was first formulated in 1979 and has since become a position common to many post-metaphysical philosophers, including, among others, Richard Rorty. A full account of this concept can be found in the volumes *Weakening Philosophy*, ed. S. Zabala (Montreal: McGill-Queens University Press, 2007), and *Between Nihilism and Politics: The Hermeneutics of Gianni Vattimo*, ed. S. Benso and B. Schroeder (Albany: SUNY Press, 2010).

toward truth"—changes "paradigms." But just as science, so too philosophy, religion, and other disciplines constantly change paradigms, such that the older theories become different, though not incorrect; in this "postmodern" condition truth does not derive from the world "as it is," but from what is responsible for its condition, namely "effective history."[4] Given these "scientific revolutions," or, as Heidegger termed them before Kuhn, "destructions of metaphysics," thinking is no longer demonstrative, but edifying, conversational, and interpretative. What does such emancipation from metaphysics entail and why is there a predilection for weakness in "weak thought"?

Contrary to some critics of weak thought,[5] such emancipation[6] does not imply a simple refusal of metaphysics, which would inevitably produce another variation on metaphysics, but rather, as Heidegger said, a *Verwindung*, distortion or twisting, in order to distance us from its frames. This distancing does not indicate a general failure, or "weakness of thinking" as such, but instead the possibility to develop all the hermeneutic potentialities of philosophy. Hermeneutics, which today is present throughout contemporary philosophy both as the philosophy of postmodernity and as evidence that our globalized culture is characterized by a clash of interpretations, has traditionally defended the weak and thus the right of different interpretations to take place. For example, what drove both Martin Luther's philological (translation of the Bible) and Sigmund Freud's psychological (stressing unconscious mental processes) revolutions were, aside from the metaphysical impositions (ecclesiastical establishment and empirical positivism) of their epochs, the demands to read the Bible independently and to enhance

4. On the relation between "weakening" and "effective history," see chap. 3 of S. Murphy, *Effective History* (Evanston, IL: Northwestern University Press, 2010), 140–196.
5. For a critique of weak thought see the essays of Jean Grondin, Michael Luntley, and Carlo Augusto Viano in *Iris: European Journal of Philosophy and Public Debate* 2.3 (2010): 105–164.
6. Such emancipation was already evident in Marder's two previous works on Derrida and Schmitt, respectively: *The Event of The Thing: Derrida's Post-Deconstructive Realism* (Toronto: University of Toronto Press, 2009), and *Groundless Existence: The Political Ontology of Carl Schmitt* (London: Continuum, 2010).

our own psychological differences. These two examples from the history of hermeneutics indicate its penchant for the discharge—what is thrown away and left to its own devices, left behind, or outside—of metaphysics, as much as its core political motivation. As we can see, it isn't for theoretical reasons that hermeneutics defends the excluded, the discharged, or, as Walter Benjamin would say, the "losers of history," but rather because they demand emancipation, the emancipation which is missing from the frames of metaphysics, where rationality sets its own boundaries, rules, and winners. This is why, as Heidegger emphasized, "overcoming is worthy only when we think about incorporation"[7]; it is the weakness of Being that allows us to overcome metaphysics, not the other way around.

What binds together weak thought and *plant-thinking* is the political motivation of hermeneutics. Although it's the weakness of Being that allows Marder to expose a philosophy of vegetal life free from our categories, measures, and frames, such freedom would be useless if it were to imply a non-hermeneutical concept of nature. After all, nature has always been framed as a normative concept both when it came to vegetal life and when it was bearing upon existence itself, determining how we should act and be regardless of our differences. This is why, from the first pages of this book, Marder presents his study as a call to "give prominence to vegetal beings, taking care to avoid their objective description and, thereby, to preserve their alterity. . . . The challenge is to let the plants be and appear within the framework of what, from our standpoint, entails profound obscurity, which, throughout the history of Western philosophy, has been the marker of their life" (introduction). This is also why, in contrast to the previous metaphysical-philosophical accounts of vegetal life (exposed in the first part of the book), where the essence of plants is determined only by applied and imposed categories, Marder suggests (in the second part of his study) that we conceive of this essence as "radically different from everything measured in human terms" because "[the plants] not only *are* but also *exist*" (chapter 2). It

7. M. Heidegger, "Overcoming Metaphysics," in *The End of Philosophy*, trans. J. Stambaugh (New York: Harper and Row, 1973), 91.

is their existence that allows us to imagine our relation to plants not in terms of "our world" (following Heidegger's sense of the term) facing a "non-world," but in terms of an interaction of two worlds ("ours" and what Marder calls the "plant-world").

Building upon the existential consequences of Heidegger's destruction of metaphysics, his ontology is not *of* the plant but *for* the plant, and in this way it is opposed to previous metaphysical positions. More significantly, Marder's ontology of vegetal life distinguishes the Being of plants from the beings themselves, that is, from the ontic categories the metaphysical tradition has always imposed on them. This is what Heidegger called the "ontological difference" between Being and beings, in which Marder envisages the deconstruction of "the metaphysical split between the soul and the body, eliminating, in the same gesture, the classical opposition between theory and practice" (epilogue). Following both Heidegger's and Derrida's destructions of the metaphysical tradition, Marder thus demands "an infinite loosening up, a weakening of the self's boundaries, commensurate with the powerlessness (*Ohnmacht*) of the plants themselves" (chapter 4).

In responding to the end of metaphysics by weakening the ontic categories of vegetal life, Marder, far from demonstrating that they were wrong, is implying that they were constrained politically, for instance, in being exclusively regulated by the "capitalist agro-scientific complex" (epilogue), which is exploiting the plants beyond any measure that could be drawn from the environment. His new book will advocate a political emancipation from this complex, as a result of the hermeneutic political philosophy touched by the Being of plants—the Being which is not an entity, but rather "a collective being," a body of "non-totalizing assemblage of multiplicities, an inherently political space of conviviality" (chapter 2). Marder nicknames this political space "vegetal democracy," which is not very different from the ecological political initiatives the so-called marginal governments of South America are activating. Just as these environmental political initiatives are constantly discredited and undermined by capitalist-run democracies, so metaphysics opposes such uncomfortable philosophical positions as weak thought, of which this book is an outstanding example.

Acknowledgments

While writing *Plant-Thinking*, I benefited from conversations with Alexandre Franco de Sá, Santiago Zabala, Luis Garagalza, Santiago Slabodsky, Artemy Magun, Maria-Luísa Portocarrero, Marcia Cavalcante-Schuback, Adam Kovach, Enrique Bonete Perales, and Carmen Velayos Castelo. Discussions of my work on the philosophy of plants at the "Politics of the One Conference" in St. Petersburg, Russia, as well as over the course of an intensive seminar on "Vegetal Anti-Metaphysics" at the University of Coimbra, Portugal, in 2010, substantially enriched this manuscript. At Columbia University Press, the ongoing support of Wendy Lochner and the fantastic work of Christine Dunbar, Kerri Sullivan, and Susan Pensak have been vital to the project. Numerous ideas that found their place in the book have been nourished by my ongoing intellectual exchanges with Patrícia Vieira, whose thinking strives to the root of things.

Parts of chapter 1 have been published under the title "Plant-Soul: The Elusive Meanings of Vegetative Life," in *Environmental Philosophy* 8.1 (spring 2011): 83–99, and are reproduced with the permission of copyright holder. An earlier draft of chapter 2 appeared in the article

"Vegetal Anti-Metaphysics: Learning from the Plants," *Continental Philosophy Review* 44.4 (November 2011): 469–489. It appears here with the kind permission of Springer Science+Business Media B.V. An earlier draft of chapter 5 was the basis for an article, "What Is Plant-Thinking?" published in volume 25 of *Klēsis: Revue philosophique* (2013), dedicated to philosophies of nature. It is reproduced here with the permission of the copyright holder.

PLANT-THINKING

Introduction

To Encounter the Plants . . .

Of all the beings that are, presumably the most difficult to think about are living creatures, because on the one hand they are in a certain way most closely akin to us, and on the other are at the same time separated from our ek-sistent essence by an abyss.
—Martin Heidegger, "Letter on Humanism"

Ab herbis igitur que terre radicitus herent, radicem disputationis sumam.
—Adelard of Bath, *Questiones naturales*[1]

The recent explosion of philosophical interest in the "question of the animal"[2] has contributed at the same time to the growing field of environmental ethics (with approaches ranging from Tom Regan's defense of animal rights to Peter Singer's utilitarian argument for animal liberation) and to the de-centering of the metaphysical image of the human, who, as we now realize, stands in a constitutive relation to its nonhuman others.

Despite strong tendencies toward the silencing of ontological preoccupations in certain ethical considerations of animality, it would be a mistake to segregate the two sets of contributions into distinct philosophical subfields. Ontological approaches to the question of the animal carry with them significant ethical implications as much for our treatment of these non-human "others" as for the human view of and relation to ourselves, while ethical debates on this question unavoidably make basic presuppositions regarding the very being of animals and humans. It is this intersection of ethics and ontology that allows current explorations of animality in philosophy to retain their critical edge, ensuring that they neither slide into a highly speculative meta-discourse on biology nor culminate in a set of normative and, in the last instance, vacuous prescriptions. And it is at the same intersection that new and more daring demands arise: to extend the scope of ethical treatment and to address the diverse modes of being of all living beings, many of them deemed too insignificant and mundane to even deserve the appellation "others."

The above twin demand, however, is yet to be heard. If animals have suffered marginalization throughout the history of Western thought, then non-human, non-animal living beings, such as plants, have populated the margin of the margin, the zone of absolute obscurity undetectable on the radars of our conceptualities. Particularly after the scientific paradigm gained its hard-won independence from theologico-philosophical dogma in early modernity, philosophers, for the most part, refrained from problematizing vegetal life, which they entrusted to the care of botanists and, later on, geneticists, ecologists, and microbiologists. The being of plants was no longer question-worthy; it did not present itself as a problem to those who took the time to contemplate it, let alone to those who made immediate use of the fruit or the flower, the root or parts of the tree-trunk.

But where the questioning impulse is dormant, ontological chimeras and ethical monstrosities rear their heads without delay. In the absence of the will to think through the logic of vegetal life, beyond its biochemical, cellular, or micro-molecular processes and ecological patterns, philosophers readily assumed that within the broad evolutionary frame of reference, the existence of plants is less developed or less differentiated than that of their animal and human counterparts and that therefore vegetal beings are unconditionally available for unlimited use and exploitation. Such suppression of the most basic question regarding plants became the breeding ground for their ethical neglect; although—akin to us—they are living creatures, we fail to detect the slightest resemblances to our life in them and, as a consequence of this failure, routinely pass a negative judgment on their worth, as well as on the place they occupy in the modern version of the "Great Chain of Being," from which both the everyday and the scientific ways of thinking have not yet completely emancipated themselves. On the obverse of this unquestioned familiarity, plants are wholly other and foreign to us, so long as we have not yet encountered them, as it were, on their own turf—so long as we have neither let them be, flourish, and appear as they are, nor done justice to them by means of this very onto-phenomenological "letting-be."

Thus formulated, our initial task is twofold: first, to give a new prominence to vegetal life, retracing the paradigm shift that had already taken place between Aristotle's investigations of *animalia* and Theophrastus's studies of plants, and, second, to scrutinize the uncritical assumptions on the basis of which this life has been hitherto explained. And yet, critique is not a panacea here: the indiscriminate dictates of a critical, analytical method and of the reason it shores up might prove to be of little use, since they cannot help but replicate past failures by *a priori* thematizing and objectifying what invites the inquiry in the first place or, negatively put, by spurning the method that we could adopt from the plants themselves. At a still more fundamental level, therefore, the question is: How is it possible for us to encounter plants? And how can we maintain and nurture, without fetishizing it, their otherness in the course of this encounter?

Human beings have a wide array of possible approaches to the world of vegetation at their disposal. More often than not, we overlook trees, bushes, shrubs, and flowers in our everyday dealings, to the extent that these plants form the inconspicuous backdrop of our lives—especially within the context of "urban landscaping"—much like the melodies and songs that unobtrusively create the desired ambience in cafes and restaurants.[3] In this inconspicuousness, we take plants for granted, so that our practical lack of attention appropriately matches their marginalization within philosophical discourses. Curiously enough, the absolute familiarity of plants coincides with their sheer strangeness, the incapacity of humans to recognize elements of ourselves in the form of vegetal being, and, hence, the uncanny—strangely familiar—nature of our relation to them.[4] In other settings, like a farm or a field of cultivated cereals, the plants' inconspicuousness is, to be sure, modified to include the instrumental attitude that sees through and past the plants themselves to what their uses might be and that treats them as nothing but potential fuel: a source of biodiesel or a vital ingredient of human and animal nutrition. When instrumentalizing plants, we do not yet encounter them, even though their outlines become to some extent more determinate thanks to the intentional comportment on the part both of those who tend them and, less so, of those who ultimately consume them. Still, the uses to which we put vegetal beings do not exhaust what (or who) they are but, on the contrary, obfuscate enormous regions of their being.

Are we perhaps in a better position to encounter the plants themselves—for instance, sunflowers—when we do not know what to do with them, are reluctant to interfere with their complex innermost workings, and simply contemplate them as they grow in a field? When we reconstruct their luminous yellowness in memory, in imagination, or on a physical canvas, as Vincent Van Gogh did at the end of the nineteenth century, depicting, more than anything else, the ephemeral nature of the flowers? Or, when we *think* the sunflowers?

The aesthetic attitude, broadly conceived, seems to be more propitious to a nonviolent approach to plants than either their practical instrumentalization or their nominalist-conceptual integration into systems of thought. In the West, nominalism has been the prevalent method of thinking about plants, integrated into ever more detailed classificatory schemas. According to Carl Linnaeus and his famous taxonomic method, I should be satisfied that I know the sunflower if I discover that it belongs to the kingdom Plantae, order Asterales, family Asteracaeae, subfamily Helianthoideae, tribe Heliantheae, and genus

Helianthus. These names are meant to capture the essence of the plant by assigning to it an exact place in a dead, albeit highly differentiated, system that swallows up the sunflower's singularity and uniqueness. The actual sunflower turns into an example of the genus, tribe, and so forth, to which it belongs and is nothing in itself outside the intricate net of classifications wherein it is caught up.

Conceptualism—particularly that of the Hegelian variety—endeavors, on the other hand, to enliven dead systems of thought, to set them in dialectical motion. But it too loses its authority in the course of considering the living flower to be nothing but a vanishing mediator, a transitory moment in the reproduction of the genus and in the passage from inorganic world to organic life, not to mention a point of transition to the fruit in a grand teleology that justifies the thorough instrumentalization of plants for animal and human purposes. Nominalist classifications and conceptual mediations join forces in violating the flower in a move tantamount to its cognitive plucking, a detachment from the ground of its existence. What they grasp, however, is a plant already dead and dry (as though it, in the first place, germinated on the page of a herbarium), deprived of its distinctiveness, and turned into a museum artifact in the labyrinths of thought.

Needless to say, the extremes of nominalism and conceptualism do not exhaust the possibilities of thinking the sunflower. Closer to the aesthetic attitude, charged with the task of reproducing or recreating the plant in imagination (and thus partaking of the reproductive potential of vegetation itself), the resources of twentieth- and twenty-first-century philosophy have much to offer to the thinking *of* plants that arises from and perpetually returns to vegetal beings. Instead of interposing additional conceptual mediations or more detailed and thorough classifications, the aim of this thinking is to reduce, minimize, put under erasure, bracket, or parenthesize the real and ideal barriers humans have erected between themselves and plants. The resources I am referring to are drawn from hermeneutic phenomenology, deconstruction, and weak thought—all consenting, in various ways, to let beings be, to save singularities from the clasp of generalizing abstraction, and perhaps to put thought in the service of finite life.

Before mining these promising reserves of contemporary philosophy for the methodology of "plant-thinking," it is worth noting that they do not stand alone in their respectful attitude toward vegetation. Non-Western and feminist philosophies contain a wealth of venerable traditions much more attuned to the floral world than any author or mainstream current in the history of Western thought. Plotinus's exceptional theoretical attention to plants and their life (to be explored below) is perhaps itself explicable with reference to his thorough knowledge of Indian philosophy and, especially, Uppanishadic and Advaita Vedantic thought. In the Indian milieu, Jain philosophy ascribed great significance to plants, so much so that it understood them to be the fifth element comprising the universe, alongside the other classical elements: earth, water, fire, and air.[5] And, in a different context altogether, one of the leading feminist philosophers in today's Europe, Luce Irigaray, semi-poetically elaborates the intimate link between a certain version of receptive subjectivity, thinking, and the plant: "The plant will have nourished the mind which contemplates the blooming of its flower."[6]

Although there is much in these heterogeneous sources that is of undeniable value to a coherent philosophy of vegetation, in the current project I limit myself to the history and "after-history" of Western metaphysics. I do so not only because the ideological roots of both the deepening environmental crisis and the exploitation of plants are buried in treatises by some of the most emblematic representatives of this tradition, but also because, on the fringes of Western philosophy and in its aftermath, surprisingly heterodox approaches to the vegetal world have germinated. The import of the external critiques of metaphysics is undeniable. But if a hope for reversing the philosophical neglect of plants in the West and for overcoming the environmental crisis of which this neglect is a part is to stay alive, immanent (internal) criticism of the metaphysical tradition must become a *sine qua non* of any reflection on vegetal life. This is why, aside from the symptomatic sites in the history of metaphysics itself, three of the most illustrious branches of post-metaphysical thought furnish, through their oft-unintended consequences, the theoretical framework for rethinking the being of plants.

Thus: Hermeneutical phenomenology advocates the kind of description that, going back to the things themselves, interprets them from the ground up and addresses each experience from the standpoint peculiar to it, all the while guarding against any unwarranted presuppositions about its subject matter. Deconstruction exposes metaphysical violence against the material, the singular, the finite; strives to do justice to what metaphysics has suppressed, while admitting that absolute justice—as the utmost attention to singularity—is impossible; and permits us to focus on that which has been otherwise marginalized without converting the margin into a new center. Weak thought resists the tyranny of "objective" factuality and welcomes a multiplicity of interpretations, even as it takes the side of the victims of historical and metaphysical brutality.

In other words, these three traditions put forth an ethical manner of thinking that permits the entity "thought through" to thrive (1) in the way it manifests itself and relates to the world (hermeneutical phenomenology), (2) in its own self-ruination and singularity (deconstruction), and (3) in its essentially incomplete hold on existence (weak thought). Thanks to their quasi-aesthetic receptivity, they leave just enough space for the sunflower to grow without trimming it down to an object readily available for the subject's manipulation, without assigning to it extraneous purposes, and without putting undue emphasis on the knowledge of its genetic makeup or evolutionary-adaptational character. Faced with a sunflower, they would resist the easy and obvious generalization of this plant as a unitary organism and would instead joyfully submit thinking itself to the ambivalence of the flower that is both one and many, consisting as it does of a myriad small florets clustered together, yet relatively independent from each other and from the community of beings we know as *a* sunflower. Succinctly put, they create the philosophical infrastructure for our encounter with plants.

In the spirit of the three traditions I have telegraphically summarized above, *Plant-Thinking* envisages the outlines of a method drawn from the plants themselves and of a discourse rooted, as Adelard of Bath has it, in these vegetal beings. By the same token, it formulates the non-transcendental conditions of possibility for encountering plants, instead of confronting them as still-murky objects of knowledge. The participants in any encounter engage in an interactive, if not always symmetrical, relationship; and, conversely, to deny interactivity is to undercut the very prospects of an encounter. The unipolarity of reason, objectifying everything in its path, and the self-proclaimed exclusiveness of the human existential comportment, which according to Martin Heidegger separates us from other living beings "by an abyss," are two salient obstacles on the road to an ontologically and ethically sensitive relation to plants.

It follows that the eventful encounter with plants whereby we find ourselves in the greatest proximity to them without negating their otherness cannot come to pass unless we entertain the hypothesis that vegetal life is coextensive with a distinct subjectivity with which we might engage, and which engages with us more frequently than we imagine. This is not to say that human beings and plants are but examples of the underlying universal agency of Life itself; nor is it to plead for an excessive anthropomorphism, modeling the subjectivity of vegetal being on our personhood. Rather, the point is that plants are capable, in their own fashion, of accessing, influencing, and being influenced by a world that does not overlap the human *Lebenswelt* but that corresponds to the vegetal modes of dwelling on and in the earth. A counterpoint to both classical and existential idealisms, plant-thinking situates the plant at the fulcrum of its world, the elemental terrain it inhabits without laying claim to or appropriating it. Strictly speaking, Heidegger is correct in arguing that plants, along with the other non-human beings, do not *have* a world, but this not-having or non-possessiveness does not at all signify the sheer absence of what we might call "plant-world," only a different relation of vegetal beings to their environment. Whenever human beings encounter plants, two or more worlds (and temporalities) intersect: to accept this axiom is already to let plants maintain their otherness, respecting the uniqueness of their existence.

Once again, the ontological and the ethical facets of the inquiry have converged at the threshold where the outlines of the vegetal world emerge thanks to our disinclination to conflate plants with complacent things immersed and dissolved in the human environment. At the same time, the task has become more difficult than ever before: we must give prominence to plants, taking care to avoid their objective description, and thereby preserve their alterity. Unlike Jacob von Uexküll, who in his highly influential *A Foray Into the Worlds of Animals and Humans* invited the reader to "walk into unknown worlds,"[7] we will not assert an unconditional right of admission into the vegetal world, which is the world *of* and *for* plants, accessible to them. The challenge is to let plants be within the framework of what, from our standpoint, entails profound obscurity, which, throughout the history of Western philosophy, has been the marker of their life. Differently put, the idea is to allow plants to flourish on the edge or at the limit of phenomenality, of visibility, and, in some sense, of "the world."

Of the three theoretical currents at the confluence of which this book stands, phenomenology, with its usual insistence on the coming-to-light and the appearance of beings under the thematizing gaze of the transcendental Ego, may seem particularly suspect when it comes to respecting the obscurity of vegetal existence. Even so, it will prove handy here, provided that it accommodates plants' constitutive subjectivity, drastically different from that of human beings, and describes their world from the hermeneutical perspective of vegetal ontology (i.e., from the standpoint of the plant itself). How does the world appear (or not appear) to a plant? What is its relation to its world? What does it strive to, direct itself toward, or intend?

A skeptic would retort that, assuming a vegetal phenomenology nascent in these questions were even plausible, it would have to be enacted by the plants themselves—an absurd proposition, to say the least. There is little doubt that the sense of the world from the animal, let alone the vegetal, standpoint remains inaccessible to us. Nevertheless, the distance between us and other living creatures loses its static character as soon as the nominal categorial divisions between various "classes" of beings are shaken and muddled, without compromising these beings' differences and commonalities. The gap separating humans from plants may dwindle—though not altogether disappear—thanks to the discovery of traces of the latter in the former, and vice versa. (What kind of traces are these? The human body and subjectivity alike are not pure expressions of Spirit but strange archives, surfaces of inscription for the vestiges of the inorganic world, of plant growth, and of animality—all of which survive and lead a clandestine afterlife in us, as us. Just as well, past and present human intentions and projections are caught up in the fabric of plant existence, reflecting histories of cross-breeding, grafting, agricultural technologies, aesthetic representations of the flora . . .) Cast in Kantian terms, vegetal phenomenology supplies plant-thinking with a normative ideal, the ideal we might approximate but never reach, unable, as it were, to put ourselves entirely in the plants' shoes, or rather roots.

The productive ambiguity haunting the title of this volume further emphasizes the ineluctable paradoxes of vegetal ontology. "Plant-thinking" refers, in the same breath, to (1) the non-cognitive, non-ideational, and non-imagistic mode of thinking *proper to* plants (what I later call "thinking without the head"); (2) the human thinking *about* plants; (3) how human thinking is, to some extent, de-humanized and rendered plant-like, altered by its encounter with the vegetal world; and finally, (4) the ongoing symbiotic relation between this transfigured thinking and the existence of plants. A sound philosophy of vegetal life must rely on the combination of these four senses of "plant-thinking," so as not to dominate (and in dominating, distort) the target of its investigations. The chances of aggravating the abuse of plants by theorizing their existence can be minimized, if the theorists themselves expose their cogitation to the logic of vegetal life and learn from it, to the point where their thinking is ready to melt into this logic, with which admittedly it will never be identical.

No genuine encounter happens without our eventful exposure, unwilled and unplanned, to that which, or the one who, is thus encountered. "Plant-thinking" is in the first place the promise and the name of an encounter, and therefore it may be read as an invitation to abandon the familiar terrain of human and humanist thought and to meet vegetal life, if not in the place where it is, then at least halfway. Part I, "Vegetal Anti-Metaphysics," clears the philosophical ground for this event by putting in question the metaphysical constructions of the plant and of what constitutes "the ground," by showing how in its very being the plant accomplishes a lived destruction of metaphysics, and by carrying out a transvaluation of metaphysical values as they pertain to vegetation. The outcomes of the critique of metaphysics are not entirely negative, given that the contours of plant life come into view as a result of the hermeneutical multiplication of its meanings released from the reductive tendencies of metaphysics that, from Aristotle to Nietzsche, ascribed to it but a single function. Indeed, these contours get redrawn as soon as we turn back to the classical notion of "plant-soul"—a notion that, despite its metaphysical heritage, exceeds the limits of metaphysics from within, to the point of overflowing them—with its countless potentialities, valid, in distinct ways, for plants, animals, and humans.

In discussing plant-soul, it would be unforgivable obdurately to insist on the traditional metaphysical separation between the "soul" and the "body" of the plant, which is only one of the many dichotomies—self and other, depth and surface, life and death, the one and the many, and so forth—practically deconstructed in vegetal existence. The positive dimension of plant-being, as the outcome of the critique of metaphysics, will spell out an inversion of traditional valuations, valorizing the other over the self, surface over depth, and so on. More crucially still, it will incorporate the core existential attributes philosophers have tended to reserve for human beings alone. I call the elements of this budding ontology, discussed in part II, "vegetal existentiality."

The intimation that it is high time to approach the existence of plants existentially is surely scandalous within the confines of Heidegger's thought and of post-Heideggerian existentialism, let alone the humanist and anthropocentric traditions of philosophy. In technical terms, its cardinal sin is that it transgresses the boundaries, firmly set in *Being and Time*, between the categorial and the existential analytics, between the categories of readiness-to-hand and presence-at-hand and the *existentiales*, such as the moods or anxiety. The initial shock the transgression may provoke will have been absorbed already in the analysis of the vegetal soul in the preceding part of the study, and it should gradually wear off once I formulate and flesh out the central concern of "Vegetal Existentiality": What are the modes of being-in-the-world appropriate to plants? Among such modes, I concentrate on temporality, freedom, and wisdom.

Just as the soul of the plant is hardly distinguishable from its body, so its time is barely dissociable from space: its outward growth and augmentation, devoid of any final accomplishment, constitutes the "bad infinity" of vegetal temporality. To be more precise, the spatio-temporal movement of plants, nonsynchronous with human time, is directed toward and by the other (light, the changing seasons, etc.) and therefore, unfolding as a hetero-temporality, is governed by the time of the other. Seasonal variation, for its part, imposes cyclical and iterable existence on perennial plants and spells out the finitude of the annual ones.

The indissoluble connection of the plant to the time of the other mirrors its spatial rootedness in the soil, a feature responsible for its coding as the figure of unfreedom. Tragically, occidental thought conflates the most plastic form of existence with the most rigid; not only does this view disregard the ontic exuberance and uncontrollable efflorescence of vegetal life, but it also ignores this life's ontological potentialities, still working themselves out in various guises in animals and human beings—the variations that free it to be otherwise than it is. On the one hand, both colloquial and philosophical discourses associate the rooted mode of being with immobility and captivity, but, on the other, the perceived indifference of plants interlaces their freedom with human liberty in the domains of ethics, aesthetics, and religion. Despite their undeniable embeddedness in the environment, plants embody the kind of detachment human beings dream of in their own transcendent aspiration to the other, Beauty, or divinity.

The living tending of plants toward their other, the tending expressed in growth, the acquisition of nutrients, and procreation, amounts to the non-conscious intentionality of vegetal life, the cornerstone of its "sagacity." In keeping with vegetal ontology, plant-thinking practices an embodied, finite, and material expression; is wholly oriented to the other without establishing either an identity or a self-identity achievable by means of its return to itself; and stands for the impersonal, non-individuated *it thinks* underlying and subverting the ever-present synthesis of *I think*, the accompaniment of all conscious representations.

The uninterrupted connection between plant-thinking and human thought finally becomes apparent when we circle back to the figure of the vegetal soul—from which the animal and human psyches emanate and which is sublimated (and, to a significant extent, dematerialized) in them—and discover the rudiments of a living intelligence in this figure.

An encounter with plants awaits us. Far from a head-on confrontation, it will verge on what, in Portuguese, is called *desencontro*—an untranslatable word, which roughly refers either to a narrowly missed meeting, a crossing of paths that was about to happen but ended up not taking place, or to an encounter that is too improbable and was never meant to happen, or, again, to a divergence of two or more (usually human) beings, each of them existing on her or his own wavelength. To meet the plants themselves, the plants as such, is not the goal of this study, if only because, in the absence of identity, they are never "themselves" and because, resistant to idealization, they do not fit within the strict philosophical confines of the "as such." All we can hope for is to brush upon the edges of their being, which is altogether outer and exposed, and in so doing to grow past the fictitious shells of *our* identity and *our* existential ontology. Are we ready to take the initial, timid steps in the anamnesis of the vegetal heritage proper to human beings, the very forgetting of which we have all but forgotten? Whether or not we will be capable of recognizing plants, and especially ourselves, in the wake of the anamnestic encounter-*desencontro* is yet to be seen.

PART I

Vegetal Anti-Metaphysics

1. The Soul of the Plant

or, The Meanings of Vegetal Life

The very fact that the acts of the vegetative soul do not obey reason shows that they rank lowest.
—Saint Thomas Aquinas, *Summa Theologica*

Psychologists no longer discussed vegetative activities.
—Franz Brentano, *Psychology from an Empirical Standpoint*

Modern readers are likely to greet positive references to "the soul of plants" with suspicion. This is not only because it seems absurd to locate the seat of the soul (and, *mutatis mutandis*, existential possibilities) in any being other than human, but also because we have grown deeply mistrustful of the heavy metaphysical and theological baggage weighing down this paleonym. Eighteenth-century French philosopher Julien La Mettrie, famous for the book *L'homme machine* (*Man a Machine*), has encapsulated his objections to his contemporaries' revisiting of the outmoded theories of vegetal soul in a lesser-known treatise *L'homme plante* (*Man a Plant*), uncharitably describing these theories as "imaginary chimeras." "How foolish of modern thinkers to try to fill such souls again with subtle breath!" he quips. "Leave their names and their spirits in peace."[1] As rigorous philosophers, we are expected to have heeded La Mettrie's injunction and so purged our thinking of onto-theological nonsense, conjuring away the pernicious old names and spirits it had been accustomed to. Whenever possible, we ought to resort to much more neutral terms, such as "the mind" (used to translate the Latin *anima* even in certain English renditions of Saint Augustine), "subjectivity," or, again, "the psyche," which, though it is the Greek word for the soul, gains in dignity by virtue of serving as the object of study in the field of psychology. Following the admonition of La Mettrie, in twenty-first-century philosophy, the "soul," which at best finds refuge in a strictly theological discourse, is finally resting in the eternal peace afforded to it by the modern repression of the word and the thing it designated. No wonder that the ancient idea of a "vegetal soul," too, is now more implausible, unfamiliar, and eccentric than ever!

But, already in its title, articulating one of the most metaphysically loaded concepts with the least metaphysical one, the present chapter evidences an attempt to conjure up these long-buried spirits and to disturb the peace of philosophical cemeteries. Indeed, it would appear that just as the invocations of "the soul" are superfluous, if not misleading, seeing that they are redolent of an outdated *Weltanschauung*, so the philosophical treatment of flora in the age of positivist science is unnecessary and is best left to the practitioners of the specialized (ontic) discipline of botany. Both verdicts have a common root in the reductively rationalized approach to reality, which has culminated in what Max Weber has called the "disenchantment of the world," where the unquestioned priority of science goes hand in hand with a delegitimization of empirically unverifiable notions. What unites the soul and plants, the most ethereal and the most earthly, is their exclusion from the purview of respectable philosophical discourses in late modernity. It is their conjunction in this space of exclusion (or exception) that will furnish us with the point of entry into the *post-metaphysical ontology of vegetal life*, in a word, "plant-thinking."

Contemporary philosophy disengages from these two entities and, in so doing, abandons them, sets them free. Left to their own devices, each transforms the other: the plant confirms the "truth" of the soul as something, in large part, non-ideal, embodied, mortal, and this-worldly, while the soul, shared with other living entities and construed as the very figure for sharing, corroborates the vivacity of the plant in excess of a reductively conceptual grasp. Within the confines of this commerce, the elusive life of the ensouled plant cannot become a scientific object without getting irretrievably lost, transformed into dead matter, dissipated in cellular activity and in the larger anatomical (or phytotomical) units, prepared in advance for vivisection.

What is in question then, in any retrieval of "plant-soul," is the very meaning of life handed over to extreme objectification and treated as though it were a plastic image of death. At the present historical conjuncture, when the wholesale transformation of all forms of vegetation into sources of food and fuel (at any rate, into something to be burned as calories or as combustibles) proceeds at an accelerated pace, it is urgent to resist the same process in thinking and to interpret the meanings of vegetal life—its precariousness, violability, and, at the same time, its astonishing tenacity, its capacity for survival—all the while steering clear of its objective and definitive determination. Only upon completing the proposed hermeneutical exercise will we be able to gauge the ethical and political implications of our treatment and mistreatment of plants, as much as the reverberations of vegetal life in beings called "human."

The Obscurity of Vegetal Life: On Barely Perceptible Motion

In various ways, ancient Greek thinkers associated life with motion. But aren't plants defined, exactly, by their incapacity to move, by their rootedness in the soil that renders them sedentary?

We find the initial intimation that the *tendency toward immobility*, as Henri Bergson expresses it, does not exhaust the mode of being of plants in the etymology of "vegetation," which points back to the Middle Latin *vegetabilis*, meaning "growing" or "flourishing," the verbs *vegetare* ("to animate" or "to enliven") and *vegere* ("to be alive," "to be active"), and the adjective *vegetus*, denoting the qualities of vigorousness and activity. The modern word "vegetable" thus deserves a patently Hegelian admiration for the speculative nature of language that invests the same semantic unit with two opposed, if not mutually exclusive, senses. While the predominant usage of the verb "to vegetate" is negative, linked to the passivity or inactivity of animals or human beings who behave as though they were sedentary plants, its subterranean history relates it to the exact opposite of this privileged meaning: the fullness and exuberance of life, vigor, and brimming energy, the *ergon* of plant-soul. Vegetal activity encrypts itself in its modes of appearance by presenting itself in the guise of passivity, which is to say, by never presenting itself as such. The life of plants, therefore, poses a special challenge before hermeneutical phenomenology, incapable of elucidating that which does not appear in the open, that which emphatically does not give itself. It is an obscure non-object: obscure, because it ineluctably withdraws, flees from sight and from rigorous interpretation; non-object, because it works outside, before, and beyond all subjective considerations and representations.

What are some of the markers of this vegetal self-encryption? Despite its apparent immobility, the plant exhibits three out of four types of movement Aristotle enumerates in *De anima*, in that it can move by altering its state, by growing, and by decaying, though not by changing its position (406a14–17). Aristotle immediately adds that "if then the soul moves, it must have one, or more than one, of all of these kinds of movement," thereby readying the theoretical space for a formal understanding of vegetal soul. It is astounding that plants are capable of motion if one identifies movement only with change of positions in space, a presupposition analogous to the modern reduction of Aristotle's fourfold theory of causality to efficient causes alone. That the plant "moves," in ways appropriate to its being, and that it is ensouled, harboring a psyche fit for its mode of living, is one and the same insight. Still alive in Johann Gottlieb Fichte, who refers to the soul of plants as "the first principle of movement in nature," albeit a principle of movement that is entirely passive, driven from the outside,[2] this idea has become completely opaque to contemporary consciousness, out of touch with the ontology of vegetal existence. Such then is the first, though by far not the definitive, meaning of plant life: a certain pace and rhythm of movement, which we customarily disregard, since it is too subtle for our cognitive and perceptual apparatuses to register in an everyday setting, and with which the tempo of our own lives is usually out of sync.

Notwithstanding the violence Aristotle's thought has unleashed against plants, his texts are fertile reserves for all those who wish to elucidate the non-metaphysical aspects of vegetal being. Among several definitions of the soul the Greek thinker provides in *De anima*, the most concise is that the soul is "the principle of animal life," *arkhē ton zōon* (402a8). It is the *arkhē* of animal life in the sense of acting as its *first* manifestation and as an *authority* that organizes and commands its further development, guiding it, in the words of Plotinus, "without effort or noise" toward its ownmost flourishing.[3] But doesn't this definition, consistent with the Aristotelian *entelechy*, deny the possibility of plant-soul by decisively locating the psyche in the sphere of animality? In its aftermath, the price for the continuing insistence on something like a vegetal psyche is the blurring of the distinction between the categories of plants and animals, a subsumption of both under the heading of "animal life," which is to say, a clandestine zoologization of the plant.

Or, is it the case that the plant has already wreaked havoc and anarchy in the metaphysical hierarchy by usurping an *arkhē* that does not rightfully belong to it but is proper to the animal? Aristotle, for instance, transgresses conceptual differentiations when he characterizes both plants and animals as "living things." But where the qualitative distinction is absent, a quantitative one takes effect, so that plants are said to be deficient and to have a weaker purchase on life than animals. To conclude that plants are defective animals is still to grant to them the rudiments of the soul but also, at the same time, to subjugate these lesser ensouled beings to those in whom the principle of life expresses itself with clarity and strength. We will be justified in holding this most obvious solution to the philosophical-taxonomical problem of the principle of vitality responsible for the devaluation of vegetal life and the transformation of the plants themselves into raw materials for animal and human consumption, a "standing reserve," on which we unreflectively draw in order to satisfy our needs.

With the view to restoring the orderliness of metaphysical categories, the life of plants becomes a matter of degree: as living things, they presumably share more with inanimate materials than with other living beings. The first manifestation of life, antecedent to its formalized "principle," is simultaneously the most reified. Assuming that the plant is an animal, it is a deficient, impassive, and insensitive one, unable to change its position in space. "Plants seem to live," writes Aristotle, "without sharing [*metekhonta*] in locomotion or in perception" (*De anima* 410b23–24). Their non-participation in the acts of locomotion and perception casts their life in the uncertain terms of a mere appearance, a matter of seeming: they only "seem to live." But even this denigration contains an unexpected promise for the post-metaphysical ontology of plants. Denied the status of the first principle, vegetal life is not identical to the underived and hence fictitious pure origin of vitality but, on the contrary, signifies whatever remains after the subtraction of the potentialities unique to other genera of the soul. After we strip life of all its recognizable features, vegetal beings go on living; plant-soul is the remains of the psyche reduced to its non-human and non-animal modality. It is life in its an-archic bareness, inferred from the fact that it persists in the absence of the signature features of animal vivacity, and it is a source of meaning, which is similarly bare, non-anthropocentric, and yet ontologically vibrant. In a word, life as survival.

The privative description of the life of plants—which, due to their proximity to inanimate or inorganic things, are even poorer in the world (i.e., more purely passive) than Heidegger's animals—is surely a reaction of metaphysical thought to the vegetal exuberance that escapes capture and taming by philosophical conceptuality. The excessive proliferation of plants (for instance, in the density of the jungle) surpasses the frames of philosophy incapable of encompassing this immoderate and immeasurable production and reproduction of life. Psychoanalytically speaking, the resourcelessness of a thought that confronts vegetation is here projected onto the very object that castrates metaphysics, spiriting the desired conceptual clarity away. Thus, in the discussion of "having" (*hexis*)—prospectively illuminating Heidegger's "fundamental concepts of metaphysics"—Aristotle cites the plant, said to be "deprived" of eyes (*Metaphysics* 1022b24), as the paradigmatic example of lack and not-having. Is this deprivation an irreparable flaw or a sign of the plant's quasi-divinity, given that in negative theology, too, we know God solely through a set of negative attributes as what He is not? And how is it to be reconciled with Theophrastus's assertions with respect to the potency of vegetal life that "has the power of growth in all its parts, inasmuch as it has life in all its parts" (*Enquiry Into Plants* 1.1.3–4)? What if, like love for ancient Greeks, the complex and ambivalent image of the plant, as much as plant-thinking itself, were a child of plentitude and lack, at once of the greatest resourcefulness and the most drastic destitution of all?

With a few notable exceptions, the exuberance of vegetal life has gone largely unrecognized in Western philosophy. Pseudo-Aristotle (most likely Nicolaus of Damascus) will intensify, in *De plantis*, the language of privation, daring to attribute to plants a lifeless soul: "But the plant does not belong to the class which has no soul, because there is some part of the soul [*meros psukhēs*] in it, but the plant is not a living creature [*zōon*], because there is no feeling in it" (316a37–40). The author of *De plantis* has carried the reduction of life to its logical extreme, where shreds of the non-animal and inanimate soul persist in the plant. It is no longer a living thing but "an incomplete thing," *ateles pragma* (316b6), something that is even less than a thing, something that awaits completion in its being productively destroyed, utilized for higher human ends of nourishment, energy generation, and sheltering. To be a plant, in the scheme of *De plantis*, is to be ontologically defective due to the position of vegetal beings close to the bottom of the teleological ladder, but also because they do not fully fit the main metaphysical categories, in this case of the thing or the animal. Inanimate things are, on this view, still superior to plants because, unlike the latter, they fully correspond to their thingly essence. The fault of the plant, therefore, hinges on the fact that it is a thing that has overstepped the confines of thinghood (but aren't all things, including the supposedly inanimate ones, uncontainable within the framework of their idealized identities?) without as yet rising to the next fully defined plane of metaphysics.[4]

What in the English translation of the text reads as the "incompletion" (*ateles*) of the plant is, likewise, its purposelessness, listlessness, lack of goal or *telos*, attributable to its non-correspondence to the relevant parts of the metaphysical paradigm. Those familiar with ancient Greek philosophy will find such designation puzzling, to say the least. Is the plant without *telos* excused from participating in the teleological scheme of metaphysical ontology? If so, is it expelled from the realm of Being? How can we reconcile such a blatant assertion of vegetal contingency with the orderly universe of the ancients? Be this as it may, it is worthwhile to examine both semantic inflections of incompletion explaining the purported defectiveness of plants, especially since they stand at the epicenter of the systematic devaluation of vegetal life in Western thought.

If incompletion means open-endedness, then vegetal growth fully satisfies this rendition of *ateles*, in that it knows neither an inherent end, nor a limit, nor a sense of measure and moderation; which is just another way of saying that it is monstrous and that equally monstrous and unbounded is the thought germinating in it. The life of a plant, metonymically associated with its growth (to say "violets grow in my garden" actually means "violets live in my garden"), is a pure proliferation bereft of a sense of closure.

We will have an occasion to revisit this construal of vegetal life as an increase of life when considering it through the double lens of Aristotle's "capacities" of the vegetal soul and Friedrich Nietzsche's will to power. For now, another permutation of limitless plant growth in nineteenth-century German philosophy is particularly relevant, namely G. W. F. Hegel's critique of bad infinity as a series that does not come to completion in a totality. Implicit in the second part of the *Encyclopaedia*, where Hegel presents his dialectical philosophy of nature, is the conclusion that the linearity of vegetal growth and the plant's constitutive failure to return to itself prevent it from having anything like a soul. Self-relation and self-reference form "a circle within the soul which holds itself aloof from its inorganic nature. But, as the plant is not such a self, it lacks the inwardness which would be free."[5] The life of a plant is limited to its outward extension, itself unlimited by anything but the environmental conditions: the amount of sunlight, the moistness of the soil, and so forth. In dialectical geometry, the plant thus finds its schematic representation in the incompletion of the line tending to (bad) infinity without closing unto itself in the circularity of a return; growth dooms the plant to strive toward exteriority without establishing any sort of inwardness, a quality Hegel associates with the soul. (We might say, somewhat ironically and with a nod of acknowledgment to Nietzsche's theory of experience and memory as a kind of indigestion, that the notion of the soul as interiority is itself an offshoot of animal physiology. The processing of energy takes place within the entrails of the animal, the inverse of the "superficial" capture of sunlight by the foliage of plants. Psychic interiority is an idealized image of the digestive tract, and what it holds inside may not be too different from the contents of the latter.) The contrast between the ancient idea of *psukhē* as an active principle of life and the modern view that necessarily ascribes to it a free space of interiority, set apart from "inorganic nature," could not be any starker. But despite this major difference, thinkers from Aristotle to Hegel have agreed upon the deficiency of linear growth, as compared to the completion of a circle, celebrated by the ancient Greek thinker both with regard to the highest perfection of thought-thinking itself and in reference to a lower capacity for self-feeling proper to the animal soul. Without a doubt, their consensus has had a negative impact on the value of vegetal life.

Plant growth is also seen as purposeless because the vegetal soul does not attain to any higher capacities other than those of endless nourishment and propagation. Having been exempted from the logic of means and ends, it may reach completion only from the external standpoint of those who will impose *their* ends onto these essentially goalless living things. The ensuing instrumental approach to plants synthesizes in itself the rationale for deforestation and the defense of forests as "the lungs of the planet," seeing that both arguments fail to take into account vegetal life *as* life, aside from the external ends it might be called to serve.

Aristotle himself would have objected to such an unabashedly instrumentalizing treatment of any ensouled being. To him, the soul is the first principle *as well as* the final cause, which is to say that "in living creatures the soul supplies such a purpose [*telos*], and this is in accordance with nature, for all natural bodies are instruments of the soul [*psukhēs organa*]; and just as is the case with the bodies of animals, so with those of plants. This shows that they exist for the sake of the soul" (*De anima* 415b16–21). The body of a plant exists for the sake of its soul (therefore, for itself), not for *our* sake. As an instrument or an organ, it is that in which the soul sets itself to work (*ergon*) and that in which it accomplishes, with more or less excellence (*arētē*), the activities for which it is fit—here, the acts of generation, growth, and nutrition.

Were we to invoke a hierarchical gradation of ends in the Aristotelian teleology and to suggest that the final purpose of plants is not exactly "final," since they are situated near the bottom of the teleological ladder, such an argument would still not justify the dialectical destruction, (or, literally, the consumption and the consummation) of the lower ends in the transition to the higher. This justification is possible only if we willfully forget, as Hegel does, about the existence of the vegetal soul, thereby reducing the plant to sheer materiality, to the case in point of spiritless and "self-less" nature. As a consequence of this forgetfulness or this repression, dialecticians will rationalize the enabling destruction of the plant's body for the sake of Spirit, as yet separate from this uninspired corporeality: "The silent essence of self-less Nature in its fruits . . . offers itself to life that has a self-like nature. In its usefulness as food and drink it reaches its highest perfection; for in this it is the possibility of a higher existence and comes into contact with spiritual reality."[6]

The life of Spirit permeates the body of the nourishing plant and elevates it on the condition that it jettison its material independence from the subject of desire and undergo a kind of productive destruction in the process of consumption. The notion of a vegetal soul becomes dialectically plausible when plants, exemplifying the rest of organic and inorganic nature, have been fully appropriated by Spirit, have shed the last vestiges of their immediate existence, and have become ennobled as a result of this spiritual instrumentalization. Weeds, of course, must be devoid of Spirit, seeing that they stand in the way of cultivating activities that render "self-less Nature" useful. As for the so-called wilderness, it occupies an ambiguous position within Hegel's system, in that it inspires human imagination and holds potentially consumable resources but does not as such attain "the highest perfection" of an apple orchard.

In a cultivated and consumable plant, Spirit will finally recognize itself as to some extent actual. But will nothing other than the plant's productive destruction trigger this self-recognition of Spirit in the world of vegetation? Aristotle gives us the tools necessary to envision an alternative and nonviolent approach, though, admittedly, its effectiveness is rather limited. For the Greek thinker, no *teloi*, high or low, would have been accomplished had the vegetal soul not set itself to work in the body of plants and, to a significant extent, in our bodies *before* any other "spiritual" interventions. It is questionable, for instance, whether the sensory and cognitive capacities of the psyche, which in human beings have been superadded to the vegetal soul, are anything but an outgrowth, an excrescence, or a variation of the latter. The sensitivity of the roots seeking moisture in the dark of the soil, the antennae of a snail probing the way ahead, and human ideas or representations we project, casting them in front of ourselves, are not as dissimilar from one another as we tend to think. Assuming then that the "higher" part of the soul is based upon, or better yet emanates from, the "lower," what does it inherit from its progenitor? How, that is, do human beings derive their identity from their inconspicuous vegetal other? In one shape or another, these will be the focal questions of *Plant-Thinking*.

We began by formulating vegetal vitality as a riddle buried in the folds of Western metaphysics. The crude solution to the problem of plant life, interpreted as qualitatively weak and as verging on inanimate existence, forces this life into retreat, puts it on the run, and so increases the distance between philosophy and vegetation. From the vantage point of Aristotelianism, the occult nature of plant life is the result of its relatively imperceptible types of movement: change of state, growth, and decay. Saint Thomas Aquinas has Aristotle's typology in mind when he writes in *Summa Theologica* that "life in plants is hidden [*vita in plantis est occulta*], since they lack sense and local motion, by which the animate and the inanimate are chiefly discerned" (q. 69, art. 2). Those features that vegetation shares with inanimate things, namely the lack of sense and locomotion, obfuscate its life processes, camouflaging vitality behind a façade of death and throwing into disarray habitual differentiations between the animate and the inanimate. Soulless yet living, the plant seems to muddle conceptual distinctions and to defy all established indexes for discerning different classes of beings in keeping with the metaphysical logic of "either/or."

Prior to Saint Thomas, the author of *De plantis* similarly oscillated between a denial that plants were living beings and an affirmation of the obscurity of their life. Animal life transpires in the open, presents itself as it is, shines forth as a phenomenon (*phanera*), and appears to be plain and obvious (*prodelos*). Vegetal life, conversely, is inaccessible, encrypted (*kekrummene*), and unapparent (*emphanes*) (815a10–13). Its movements are so subtle that it is easy to mistake a dormant tree in the winter for dead wood, the archetype of inert matter. It follows that to raise the question of vegetal life phenomenologically, by chasing it out of its concealment and by shedding light onto it, is already to violate this life, to overlook its non-phenomenality. And conversely, to get in touch with the existence of plants one must acquire a taste for the concealed and the withdrawn, including the various meanings of this existence that are equally elusive and inexhaustible.

The fugal, fugitive mode of being, responsible for the unapparent character of vegetal life, replicates the activity of *phusis* itself, which, according to the famous Heraclitean fragment 123, "loves to hide," *kryptesthai philei*. The cryptic life of plants stands for the synecdoche of self-veiling nature—for *phusis*, which, in its Greek derivation from the root *phuo-* and the verb *phuein* ("to generate," "to grow out," or "to bring forth"), alludes to the world of vegetation and the plant (*phutō*).[7]

The parallel between nature as a whole and the plant is a promising beginning for the philosophy of vegetal life. On Heidegger's reading, the emergence of nature, or nature *as* emergence, as a surge into being, is at the same time its retreat, a giving withdrawal and an inexhaustible generosity.[8] *Phusis*, with its pendular movement of dis-closure, revelation and concealment, is yet another—not fully ontologized—name for being, which is and is not identical with everything that is *in* being and the meaning of which is lost in every attempt to name it. Life and the soul, similarly, first emerge in the plant only to retreat from it following its merciless reification, the inflation of its thingly dimension, and the forgetting of its ontological makeup. But while Heidegger attributes a positive function to the negative moment of being's withdrawal, casting it in terms of the indispensable underside of truth as un-concealment (*a-letheia*), the ancient insights on the encryption of life in the plant give rise to its mystifying fetishization.

Fetishism, *nota bene*, is a dangerous but not unavoidable supplement to the ontological approach to vegetal life. For the fetishist and animist mentalities, although plants bear resemblance to mere things, they engender a mysterious excess over other inanimate entities, the excess that, inexplicable and miraculous within a reified order, is treated as worthy of veneration. The early religious fertility cults are of course the most unsublimated version of venerating something non-thingly within the thing, something that makes it alive and that does not quite fit into the fully substantialized, rigid, and concrete panorama of reality. Wrapped in the covers of myth, vegetal life turns all the more numinous and obscure, so that its meanings are completely withdrawn, made unapparent and indiscernible, paving the way for the projection of human purposes and goals onto it. Whereas the complete phenomenalization of life leaves nothing to interpretation, because everything has been placed in the open, the plants' becoming-noumenal likewise forecloses hermeneutical ventures, insofar as it reduces the meaning of vegetal life to pure meaninglessness. As plants testify in their own manner, life, onto-phenomenologically conceived, is the process of coming to light that is not entirely victorious over obscurity. Symbolically then vegetal existence could be seen as a metaphor for vivacity itself: the germination of a plant striving toward the light of the sun happens simultaneously with its roots burrowing ever deeper into the darkness of the earth: "While the 'plant' sprouts, emerges, and extends itself into the open, it simultaneously goes back into its roots in that it fixes them in the closed and takes its stand. The self-unfolding is inherently a going-back-into-itself."[9] The plant's (and thought's) deracination and total exposure to light make it perish, as does its isolation from the sun's luminous warmth.

The fragile balance of light and darkness, of the open and the closed, required for the plant's biological life is equally applicable to its persistence as a living figuration of thought; if we are to "think the plants," we must not shy away from darkness and obscurity, even as we let them appear in their own light, the one emanating from their own kind of being. In remarking that "to establish [the plant life's] existence requires considerable research" (815a13–14), pseudo-Aristotle appeals to what we may call a "hermeneutics of vegetal life" as a way of tearing it out of concealment without determining its meaning once and for all. If it is to be effective, such hermeneutics must on the one hand precipitate a critique of philosophy that has thus far forced the life of plants into retreat, exacerbating the ownmost tendency of vegetal vitality, and on the other sustain a delicate equilibrium between the extremes of fetishistic obscurantism, which denies the very possibility of meaning, and a scientific-phenomenological elucidation of that which is withdrawn. A critique of philosophy—or more precisely, deconstruction of the metaphysical representations of plants—is the preparatory work needed for the hermeneutics of vegetal life to flourish in the conceptual space of semi-obscurity conducive to this life.

Abstract as it might seem, the philosophical denegation of vegetal existence has had palpable effects on the human approach to natural environment, so that, for example, the woods are treated as nothing more than wood, a mass of lumber "produced" in a gigantic and infinitely stocked factory of planetary proportions. This example is not accidental, given that the concept of *matter* arose in Aristotle's thought by way of adopting the everyday word for timber, *hulē*, for rigorously philosophical purposes. But while Aristotle still imbued *hulē* with the dignity of the material cause, for the modern scientific consciousness it designates nothing more than the shapeless stuff awaiting an external imposition of form. In light of this conceptual prehistory, all that is required is to project the impoverished notion of matter back onto its pre-philosophical source (*hulē* or timber) and so to confirm, in a vicious circle, that the woods are wood awaiting its elevation—as Hegel would have it—or the sublation of its immediate existence into the form of a house, a page in a book, or logs in the fireplace. For, and one should keep this in mind, essentially "incomplete" things become what they are only when they are on the verge of no longer being.

In response to the regrettable identification of vegetal life with mute and inert matter, it is imperative to make the first, tentative steps toward acknowledging that this elusive vitality is the embodied limit of the metaphysical grasp and is therefore unapparent, hidden, and above all encrypted, *from the standpoint of metaphysics* that unwittingly sides with ancient animism. Needless to say, the practical outcomes of considering the plant as one of the signposts of philosophy's finitude, situated both below the threshold of metaphysical understanding and at the much more positive limits of vegetal hermeneutics, will include a drastically different comportment toward the environment, which will no longer be perceived as a collection of natural resources and raw materials managed, more or less effectively, by human beings. And since plants are the synecdoches of nature as a whole, their philosophical defense bears upon all of *phusis*, without running the risk of replicating the abstract, general, and indifferent metaphysical thinking enamored with totalities, such as nature or indeed the environment.

There is, however, an additional paradox in the assertion that the life of plants is "hidden." For Aristotle, as for Hegel, plants are essentially superficial, and this makes certain botanical sense, given that they strive to a maximization of their surfaces in order to capture as much solar energy as possible. At the same time, unaware of the exchange of gases between plants and the atmosphere, the Greek philosopher considered their soul to be incapable of breathing (*pneuma*)—an ethereal process synonymous with the soul and one that bespoke a certain hiddenness of the organ of respiration, the lungs.[10] In the same spirit, the German thinker postulated an immediate identity between the inner life of the plant and its outer vitality. If plants have something like a soul, they wear it on their sleeves, so to speak, since "the plant's vitality in general . . . does not exist as a state distinct from the plant's inner life."[11] In the face of these imputations of absolute superficiality to plants, how is it possible that something would be hidden there where the dimension of depth is absent? And what is the relation between this sort of hiddenness—call it "superficial hiddenness"—and the withdrawal of the human soul to subjective interiority—call it "profound hiddenness"?

A comparable puzzle lies at the core of Heidegger's ontological reading of phenomenology, where being is encrypted not in the deepest recesses of an entity (as it is in Hegel's philosophy, before the dialectical mediation of being's essence with its outward appearances) but right on the superficies of the ontic. Ontico-ontological difference is in this sense superficial. Hermeneutics realizes the value of such superficiality: rather than track down profound meaning, in the manner of an archeology of knowledge, it renders explicit what has been always already vaguely "pre-understood," what has been right on the surface of things, too close to us to be considered questionable. What is hidden and distant from us is the most obvious, that which is taken for granted and unnoticed because of its intimate familiarity; it is being itself. Instead of concealing a deeply buried secret, the encryption of vegetal life refers to this life's unquestioned obviousness, to the soul of plants that is so close to us that it to a large extent and unbeknownst to us constitutes human beings.

Precisely with reference to the "breathing" of the plant and on the brink of making a transition to the philosophy of animality, Hegel intensifies the paradox and admits that this "process is obscure because of the sealed reticence of the plant [*verschlossenen Ansichhaltens der Pflanze*]."[12] A closed reserve, the plant, whose negativity is now intensified, holds back, keeps to itself, withholds its teaching—as Socrates notes in *Phaedrus*: "The country and the trees teach me nothing, whereas the men of the city do teach me" (230d)—and passively resists all efforts at comprehending it. Unlike an animal, the plant has no voice (this explains its reticence), and it is incapable of spontaneously choosing its place by exercising the freedom of self-movement (which justifies its sealed character). Indifferent to the distinction between the inner and the outer, it is literally locked in itself, but in such a way that it merges with the external environment, to which it is completely beholden. In other words, it is absolutely other to itself and, as such, transcends the relative and reciprocal distinction between sameness and otherness. It poses an obstacle on the path of metaphysical thought that traffics solely in identities and self-identical units and that regards all else as obscure, sealed, and reticent. But at the same time it is this reticence of the plant that Spirit exploits in speaking *for* the sealed and obscure entity, in feigning to become its mouthpiece, and filling in the lacuna of non-identity, or in the Plotinian vernacular, the "otherness" of vegetal desire.[13]

Like nature, with which it stands in a synecdochic relation and which is only initially other to Spirit, the plant undergoes spiritualization and elevation at the price of its productive destruction wrought by *Aufhebung,* the dialectical sublation.[14] Spirit interposes itself into the place of the vegetal soul it has refused to recognize. In so doing, it claims the absolute right of appropriation over the mute body of the plant, sublimated, for example, into the divine body, the Eucharistic blood and flesh of Christ, as a consequence of its concrete negation in the humanly (spiritually) controlled processes of fermentation: the transformation of grapes into wine or "spirits" and of wheat into bread. Through the sanctified human activities of cultivating certain kinds of plants and transforming them into edible or drinkable substances (here, I repeat, we are dealing with a very telling example), the subaltern plant, itself incapable of speech, is represented by and commences to speak with more than one voice and in more than one tongue: it comes to ventriloquize at once the voice of Reason and that of Revelation . . . and so ceases to be a plant.

When Spirit speaks for and misrepresents the plant, it does not thereby break the sealed reserve of vegetal life. It would be plausible, in the Heideggerian vein, to attribute the reticence of this life to its provenance, to the originary vivacity, ontologically understood as the event of propriation (*Ereignis*), the very self-giving of being, that withdraws and withholds itself from every human attempt to appropriate it. This conclusion would be in line with Aristotle's earlier insistence on the original status of plant-soul, "a kind of first principle in plants [*phutois psukhē arkhē*]" (*De anima* 411b28–29). We might notice, nevertheless, that the Aristotelian-Heideggerian hypothesis loses sight of a great deal of inauthenticity implicit in this impure origin of life—the fragility or, as Hegel puts it less kindly, the "feebleness" of vegetal vitality.[15] Life's principle is still too weak in the plant, the soul of which is neither differentiated in its capacities nor separate enough from the exteriority of its environment. But what is weakness for metaphysics marshals a strength of its own,[16] both in the sense of passive resistance it offers to the hegemonic thinking of identity and in the sense of its independence from the fiction of a strong unitary origin. The botanical event of propriation is necessarily that of primal ex-propriation, either of the plant by itself, i.e., by the absence of its self-identity, or by animal or human beings.

Among the ancients, Plotinus is the thinker most attuned to the originary "impurity" of plant-soul, which he variously describes as "a shadow of the soul [*skian psukhēs*]" (4.4.18.7), and as a "kind of echo of the soul" (4.4.22.2). The conventional interpretation of the shadow and the echo as derivative from original sights and sounds buttresses the Plotinian speculation that the living and ensouled earth itself is responsible for the germination of the seed hidden in it and that the earth therefore stands closer to the origin of life than does the vegetation it nourishes and supports: the vitality of plants echoes the more intense life of the earth. At this point on the quest for a purer origin, ancient animism is in collusion with metaphysics. And yet there is an alternative way to inherit the suggestive formulations of Plotinus, to read them against the grain by locating the more or less obscure repetition and similitude—the shadow and the echo—at the source of life produced *as* a reproduction, the origin of which is deferred *ad infinitum*. Life is an echo of itself, resonating with equal non-originarity in all living beings, incapable of ever appropriating it. The echo and the shadow of the soul are not its pale copies but the most faithful figurations of the psyche in the incessant process of becoming. They are especially pertinent to plant-soul, since they help maintain the precarious balance between obscurity and luminosity both in the existence of and in the theoretical elaborations on vegetal beings.

In the terms of contemporary philosophy, the echo and the shadow are traces, presences that are from the outset "impure," contaminated by absence. Somewhat closer to us, F. W. J. Schelling reiterates the Plotinian insight when he writes that "in every organization there is something *symbolic*, and every plant is, so to speak, the intertwined trace of the soul."[17] The symbolic constitution of everything, including nature, implies that the entire universe is at least potentially meaningful and that meaning is not at all separate from the life of every organization and of every organism but is coextensive with their ontological dimensions. These structures of meaning are not objectively metaphysical, immutable and pre-given, like the Book of Nature or the DNA code, awaiting their decipherment; rather, "the trace of the soul" determines symbolic constitution from the standpoint of what is so constituted, in and through the act of living itself. Phenomenologically speaking, the world becomes meaningful (or selectively illuminated) *for someone*, for a consciousness that has experiences by virtue of sense-bestowal (*Sinngebung*) positing the being of its object, or for a life lived outside the purview of consciousness. As a consequence of Schelling's intriguing idea, within the broad epigenetic conception of nature as suffused with subjectivity, plants, carrying traces of the soul, are not mere objects to be studied and classified; they are also agents in the production of meaning (a vegetal "autoproduction" of meaning without the interference of thought, in a succinct formulation of Maurice Merleau-Ponty),[18] even if this meaning is pertinent to their generative and nutritive capacities and activities alone. What appears to be meaningless and obscure *to us* becomes meaningful as soon as we try to imagine, at the edge of our imaginative capacity, the perspectives of those beings that live unconcerned with symbolic meanings. The old question about the "meaning of life" should as a result give way to questions about the meanings of *lives* (both human and non-human) that arise, practically and concretely, from the heterogeneous vivacious activities of every single creature, including a plant.

To ensure that the trace of the plant's soul is not irretrievably lost in the massive objectification of vegetal life proceeding at an accelerated pace today, in the early years of the twenty-first century, it is necessary to transpose the categories Heidegger reserved for *Dasein*, or, simply, for human existence, back onto "objective" nature. Admittedly, this transposition will not be tantamount to a direct translation, since it cannot ignore the qualitative differences between human and plant lives. Provided that the notion of the trace is taken seriously, the following questions will immediately confront us: What are the aspects of Heidegger's existential analytic that may survive their projection back onto vegetal life? How and in what shape are they going to persist? What is the sense of survival operative in this transposition? And what of the plant's soul lives on in us? I will take up these and related questions in my subsequent theorization of "vegetal existentiality" in part II of this study.

In deconstruction too the trace is a weak presence, an imprint fatefully entwined with the absence of that which left it. But it is also a synonym for survival, the continuation of a life shaken up by a rupture (trauma, for instance) portending death. The twofold question apropos of the mutual survival of plant-soul in human beings and of the qualities of *Dasein* in the world of vegetation is a part of the economy of weak presence that locates traces of the plant in the human and traces of the human in the plant. We cannot help but feel a tinge of the uncanny in the demand that we discern the constitutive vegetal otherness in ourselves and simultaneously relinquish the illusion that *Dasein* with its ontological comportment is exclusive to human beings, while all other manifestations of life are narrowly ontic. The other who (or that) bestows upon us our humanity need not be—in keeping with Aristotle's preferred points of comparison in *The Politics*—a god or a beast, the magnificently superhuman or the deplorably subhuman. It may well be the most mundane and unobtrusive instance of alterity, to which we do not (already or yet) dare to compare ourselves: the plant.

The Potentialities of Plants; or, The Vicissitudes of Nourishment

The starting point for our inquiry had to do with the basic signification of life as motion and the rather counterintuitive attribution of this sense of living to plants. Aristotle further specifies the life of the soul in terms of a capacity (*dunamis*) for at least two types of movement—growth and decay (*De anima* 412a14–15)—and for the absorption of nutrients. If life betokens "the movement implied in nutrition and decay or growth," then "plants are considered to live, for they evidently have in themselves a capacity and first principle [*dunamin kai arkhēn*] by means of which they exhibit both growth and decay in opposite directions; for they do not grow up and not down, but equally in both directions, and in every direction" (413a26–30). It will be recalled that the capacities are not superimposed upon the Aristotelian soul, which is actually inseparable from them, but that instead they denote active, dynamic tendencies, not passive features of the psyche. To be capable of something is to actively strive toward that of which one is capable, to be directed toward it with one's whole being, to find one's very being in this striving. In Edmund Husserl's appropriation of Aristotle, to be capable of . . . is to have intentionality, which is a directedness-toward something, be it the perceived, the desired, the willed, or—we might add—for a plant, light, moisture, mineral nutrients. Regardless of its content, the formal assertion that the plant is *capable of* something already endows its existence with qualities that are not entirely passive.

The *dunamis* of the vegetal soul, its capacity for growth but also for decay and the assimilation of nutrients, sets itself to work in a seemingly limitless extension in every conceivable direction, not just in a heliocentric tending toward the light. Plant life expresses itself both by means of biochemical signaling and in an incessant, wild proliferation, a becoming-spatial and a becoming-literal of intentionality. Multidirectional growth is already in and of itself the budding of dense meaning and sense—*sens* and *sentido*, meaning in French and Portuguese both "meaning" and "direction." That this *non*-conscious intentionality of the plant edges closer to the *un*conscious is obvious both within the Aristotelian scheme, where there is no "difference between slumbering without being awakened from the first day till the last of a thousand or any number of years, and living a vegetable existence" (*Eudemian Ethics* I, 1216a1–10), and to the readers of Bergson, who nevertheless recommends that the definition of the vegetable "by consciousness asleep and by insensibility" be dynamic enough to accommodate those rare instances when "vegetable cells are not so sound asleep that they cannot rouse themselves when circumstances permit or demand it."[19] It is thus possible for the life of the plant to awaken, if only for a brief moment, to come out of its obscurity, countering the tendency of animal sensibility to fall back into the torpor and immobility of the vegetable. The replacement of rigid taxonomies with fluid becomings in Bergson's work synchronizes the tendencies of distinct kinds of life, whether animal or vegetal, with the dynamic capacities of the Aristotelian soul, inexhaustible in the terms of the static "ladder of Being," wherein the notion of the soul was imprisoned in medieval philosophy. Meaning in its spatial becoming is what plants enact by exercising the capacities of their soul.

Vegetal life, with its seemingly infinite proliferation, displays an exuberance of growth and an equally spectacular decay that in their excessiveness put to work the capacities of plant-soul without ever fully actualizing or accomplishing them. Within the framework of actuality, this life is a failure, an unfinished project, but so too is human existence, unless its incompletion is positively understood from the existential point of view. To be sure, a productivist teleology may impute finality to the plant's coming to fruition, but this imputation would be alien to the living of life inexhaustible in any of its tangible "outcomes."

Although vegetal life lacks an objective end, Aristotle, like many philosophers in his footsteps, chases after its elusive first principle, the basic capacity and the unitary origin of the soul from which he would deduce all the others. According to *De anima*, the generic *dunamis* of this life is the nutritive faculty, *to threptikon*, homologous to the fundamental haptic sense in animals (in a word, touch), which is subsequently differentiated into other specific senses (413b1–10). *To threptikon*, Aristotle contends, is the precise place where the soul begins in a simple unity that will grant life to plants and to all living beings without exception. It is the minimal level of vitality that distinguishes living entities from mere things, and the plant stands right at the threshold of this distinction, given that no other capacities supplement *to threptikon* in its sphere of existence.

In a tacit allusion to Aristotle's text, Nietzsche mischievously carries the reduction of the classical capacities further, when in a fragment dated 1886–1887 he concludes, "'Nourishment'—is only derivative; the original phenomenon is: to desire to incorporate everything."[20] With this, he weighs in on the now-forgotten ancient debate that unfolded around the speculation as to whether plants experienced desire. Whereas Plato and his followers were convinced that plants could be counted among desiring beings, Aristotle vehemently denied this conclusion. Plato's indications on the subject of vegetal desire are at their most revealing in *Timaeus*, where the soul of a rooted living being (that is, of the plant *qua* an inferior animal) is thought to share "in sensations, pleasant and painful, together with desires [*epithumiōn*]," despite being incapable of self-movement (77b). The sensate, desiring vegetal soul thus already includes elements of *to phronimon*, intelligence as discernment.[21]

The unstated premise of the argument for the plausibility of vegetal desire is a supposition, which pseudo-Aristotle subsequently articulated,[22] that what is capable of receiving nourishment is subject to the feelings of hunger, craving, and satisfaction depending on whether nutrients are available at any given moment. On this view, desire (first and foremost, plant desire, to which we are also privy whenever we are hungry or thirsty) is negative, predicated on lack, and satisfiable exclusively in those brief intervals when the organism is sated. Against the background of this deficient or defective desire, the exuberance of vegetal life is but a veneer overlaying a profound absence of fulfillment, the default state of all living, hetero-affected beings reliant on something outside of themselves. But if this is so, then the plant is the most desiring being of all, precisely because it is the one most dependent on exteriority.

Should we accept it as an axiom that negativity is the essence of desire, let alone of vegetal desire, if such a thing is conceivable? Nietzsche sides with Plato in the attribution of desire to the nourished living entities, but unlike the Platonists he uncouples it from the sensations of pleasure and pain or, more broadly, from the connotations of absence and lack. The Nietzschean nutritive desire is an expression of the overflowing will to power, the pure positivity of growth and expansion where nothing is missing. Even if its object is a neutralized other incorporated into the same, its most profound source—proper to any living being nourished by assimilating the other to itself, by destroying this otherness, and by drawing its energy in the process—is the positivity of self-affirmation, an increase in strength. Having stated the issue at the highest degree of abstraction, Nietzsche implies that "higher" organisms and psychic processes have never really superseded this basic modus operandi of plant-soul. Instead, "to this mode of nutrition, as a means of making it possible, belong all so-called feelings, ideas, thoughts."[23] In an ironic amplification of Aristotelianism and Hegelianism, the vegetal capacity for nourishment, or more generally for the assimilation of alterity to the same, is gradually sublimated into ideas and thoughts that finesse and spiritualize the strategies of incorporating the other, of feeding themselves on difference, and of harnessing desire for dematerialized ends. (Think back, on the one hand, to Hegel's *Geist* and how it idealizes the nutritive principle of assimilation converted into a method for building a totality, and on the other, to Aristotle's assertion that without the nutritive faculty, the receptivity of sensation would not have been possible.) Philosophy itself becomes but the most refined and sublimated version of *to threptikon*, where the act of thinking embodies the living legacy of vegetal soul's signature capacity. Even in our highest endeavors, we remain sublimated plants.

It is thus particularly unfortunate that Nietzsche's brilliant intuition is marred by his reductive view of the plant as a vegetal manifestation of the will to power. In Heidegger's narrativization of the history of Western philosophy, Nietzsche has produced the last variation on Platonism by turning it upside down, by revaluing the highest Platonic values (for instance, the Ideas) as the lowest. The nineteenth-century thinker's name for being is "will to power," the spring of the plant's capacity for nourishment and of the desire to assimilate the other that underpins this capacity. "'Nourishment,'" Nietzsche writes, as though reinforcing the already-cited passage, "[is] only a consequence of insatiable appropriation, of the will to power."[24] Underlying the exorbitant ontic growth and decay of vegetation, as well as the ontology of plant life as a process of incessant proliferation, is the insatiable desire to appropriate the other, to grow in force. It would seem that plants act on this desire in the most literal sense, by branching out in all directions: growing in height, spreading horizontally across vast expanses, burrowing their roots deep into the Earth's crusts, and by imbibing everything from the water, the air, and the soil that surrounds them. Their "other" is the entire inorganic mineral world, the world they conquer both by spatially spreading themselves on the surface of the planet and by "digesting" mineral nutrients. This is why the jungle is Nietzsche's favorite example for the material workings of the unstoppable will to power in plants ("For what do the trees in a jungle fight each other? For 'happiness'?—*For power!*—") fighting for their place under the sun and trampling other vegetal beings in the process.[25] At the hands of Nietzsche, then, vegetal life loses its multiple semantic layers, gets torn out of its obscurity, and is reduced to little more than the conquest of inorganic elements accompanied by a struggle of plants against one another.

But this is precisely Nietzsche's error: besides projecting anthropomorphic feelings and behaviors onto plants, he includes them under the concepts of sameness and identity. He ignores the fact that in the absence of a clearly demarcated space of psychic interiority, they are incapable of incorporating anything in their souls which merge with the materiality of their bodies. The paradox is that the insatiability of nutritive desire coincides, in the plant, with the nonexistence of an autonomous self to which the other would be appropriated. In the absence of identity, the increase of power for the plant "itself" implies the augmentation of power for its "other," be it another plant or inorganic nature as a whole. Surprisingly, Hegel deserves credit for being more sensitive to this issue than Nietzsche and for proposing that the plant's "assimilation to itself of the other . . . is also a going-forth-from itself,"[26] an interiority immediately identical to the process of exteriorization. Still, for Hegel the plant's inability to establish an identity with itself by means of the other is a vice, whereas for post-metaphysical plant-thinking it is a virtue, a prerequisite for the thought of difference and non-identity incompatible with the imperialistic appropriation of the other.

From nutrition through the assimilation and appropriation of the other to the same, to the will to power—the chain of reductions to the fundamental capacity of plant-soul winds on in an infinite regress to the evanescent first principle, rendering every new term more metaphysical and abstract than the preceding one. Nietzsche explains the latest and the most vital link in the conceptual chain as a desire for the accumulation of force: "The will to accumulate force is special to the phenomena of life, to nourishment, procreation, inheritance—to society, state, custom, authority."[27] The philosopher harnesses the exuberance of vegetal life, its untamable proliferation, for a definite end, that is, the will to power that desires in the last instance the accumulation of more power (more life). Nietzsche does not entertain the hypothesis that the phenomena of life, and among these the vitality of plants, often preclude the hoarding of power. Their unique ensouled existence enjoins plants to be the passages, the outlets, or the media for the other. What if, consistent with this conclusion, the advantage of plant-soul and plant-thinking is that they let the other pass through them without detracting from the other's alterity? What if they grow so as to play this role more effectively, to welcome the other better? What if all this is accomplished thanks to the essential incompletion of linear growth that does not return to itself but is, from the very outset, other to itself? And what if, finally, this inherent respect for alterity spelled out the multiple meanings of vegetal life?

On "The Common": Modes of Living and the Shared Soul

The breaking point in Nietzsche's meditations on the plant is his analysis of its synthetic unity into a multiplicity of growths resisting the drive toward accumulation and totalization. Inquiring into something we may recognize as the non-conscious intentionality of vegetal life, its directedness-toward . . . —"What does a plant strive for?"—Nietzsche responds: "—but here we have already invented a false unity which does not exist; the fact of a millionfold growth with individual and semi-individual initiatives is concealed and denied if we begin by positing a crude unity 'plant.'"[28] The striving of vegetal beings is not a simple unidirectional effort; the non-conscious intentionality of plants and plant parts (which, like the florets within the sunflower, are not at all distinguishable from the vegetal "whole") is hopelessly dispersed and disseminated. In these lines, then, Nietzsche de-idealizes the plant and thereby liberates the difference imprisoned in this conceptual unit, just as roughly a century after him Jacques Derrida would release packs of heterogeneous animals from the constraints of "the animal"[29] and multiple things from the identitarian stricture of "the thing itself."[30] Plant-thinking is the thought of intentionality's un-synthesizable dispersion: whereas Hegel's "plant-life . . . begins where the vital principle gathers itself into a point,"[31] Nietzsche's vegetal vitality commences with the atomic fission of the unitary principle into an infinite number of points. Plant-thinking starts with the explosion of identity.

How does the material analysis of crude vegetal unities into subconceptual multiplicities bear upon plant-soul? From Aristotle to Nietzsche, philosophers have depicted the vegetal psyche as a loosely organized conglomeration of souls, a synthetic assemblage where the unity of the whole is only provisional. Life itself is lived primarily in and as dispersion—Aristotle was already acutely aware of this insight, against which he struggled in his thought, dedicating all his energy to finding a formula that would permit life to return to itself, to be gathered in itself. Despite furnishing the indelible image of *theoreia* in response to the demand to concentrate the life of the mind in a coherent totality, the Greek philosopher grants that biological life is necessarily dispersed, for instance when he takes the empirical observation that once a twig is separated from the mother plant, it will become a new plant, to mean, on a metaphysical plane that parallels certain strands in Jain philosophy, that each vegetal being potentially has more than one soul: "For just as in the case of plants some parts clearly live when divided and separated from each other, so that the soul in them appears to be one in actuality in each whole plant, but potentially more than one [*dunamei de pleionon*]" (*De anima* 413b15–20). The analogy crops up in the course of discussing the faculties of the soul (the nutritive, the sensitive, and the cogitative) and tackling the challenge of comprehending parts of the soul based on divisions in space. Aristotle deems the problem to be easily resolvable when it comes to plants and certain animals, like worms, that continue to live even after being cut in half. In these cases, the unity of plant-soul is in actuality a mere appearance concealing the potential proliferation of souls, manifold and divisible. Plant-soul is in and of itself a conglomerate of plant-souls: both one and many.

The infinite divisibility of the nutritive soul, as well as of certain sensory animal souls, makes it approximate the body, defined through this feature of extension as much in the writings of Aristotle as in Descartes. So intimately is the extended vegetal psyche bound to the body it animates that its nature is barely distinguishable from that of corporeal entities. The plant's life is indissociable from the finitude and materiality of its soul, and this is why Aristotle concludes that this soul is perishable, subject to degeneration and decay, in contrast to the "soul of another genus," *psukhēs genos heteron*, the mind and the immortal faculty of thinking (413b25–27). The division in the soul between the divisible and the indivisible complicates a straightforward opposition between the simple psychic unity and the composite character of the body: the synthetic—and, therefore, prone to being analyzed or broken down—structure of the soul belongs to plants as much as to animals, as Fichte later explained in his theory of the animal psyche comprised of "a system of plant-souls."[32] Freudian psychoanalysis crosses the next frontier when it postulates the *a priori* divisibility of the psyche into the conscious and the unconscious, itself differentiated into a network of traces. The object of *psycho*-analysis, wherein we might detect a vegetal approach to the psyche, is no longer "a soul of another genus" but an extended psychic thing entwined with the body itself—a somatic, and thus divisible, soul akin to that of a plant.[33] Post-metaphysical thought, such as that of psychoanalysis, no longer believes in the fiction of the indivisible and immortal soul "of another genus." Psychic divisibility becomes the destiny of humanity that, perhaps without knowing it, sets for itself an infinite task: that of recovering its vegetal heritage.

Whether it puts itself to work in plants or in human beings, the divisible vegetal psyche does not prevent the formation of fleeting collectivities, or loose assemblages, that, at the extreme, give off the appearance of independent organisms and monolithic social or political entities. Nietzsche's reduction of the unity "plant" to proliferating multiplicities reaches out, by the same stroke, to the Hegelian determination of the plant in terms of a "difference in itself" or an inner "dispersion into a multiplicity of . . . forms"[34] and, still further back, to the Aristotelian plant-soul, where the dynamic unity of what has been dispersed defines this soul's nutritive capacity. In an effort to envision this unity in flux, from the perspective of becoming rather than being, Nietzsche resorts to a peculiar elucidation of "life." "A multiplicity of forces," he writes, "connected by a common mode of nutrition, we call 'life.'"[35]

There is then a way to bring back together multiple "individual and semi-individual initiatives" of growth that had been indiscriminately absorbed into the concept of the plant without homogenizing them, without losing their singularity. Life in Nietzsche's rendition is a trajectory, temporarily gathering the diverse; as he specifies later on: "'Life' would be defined as an enduring form of processes of the establishment of force, in which the *different contenders grow unequally*."[36] But what is it that allows this new term to succeed, almost magically or alchemically, there where other unitary notions have failed? Why does difference (here, the dispersion of vegetal growths) escape unscathed its incorporation into "life"? And does the act of living *necessarily* trigger various inequalities among "different contenders"?

The common thread tying together the multiple quanta of force in the first definition of life is the nutritive capacity, the mainstay of the vegetal soul. What is significant, I believe, is that in accounting for this red thread, Nietzsche privileges not so much nutrition itself as its function as a common mode combining a multiplicity of forces. What is the sense of this commonality? It can imply three things: (1) that when it comes to a single living entity (a tulip, for example) the unity of its roots, stem, leaves, and flower is due to the fact that these moments of growth are traversed by a unitary network of vessels delivering nutrients to each part; (2) that different modes of nutrition mark different forms of life: plants are distinct from animals because the former draw nutrients from the soil through their roots (hence their immobility), whereas the latter devour plants and other animals (hence their mobility); or (3) that all living beings are alive, participate in the act of living, to the extent that they are able to be nourished, or share nutrition as a common mode of being. Nutritional commonality will therefore apply to parts of the same being, a group of similar beings, or all living creatures, depending on the extent of the network wherein multiple, previously disparate quantities of force temporarily come together.

The third sense of being in common Nietzsche's fragment suggests—that all living beings are basically alive thanks to being able to be nourished—is entirely consistent with Aristotle's claim in *De anima* that "the nutritive soul [*threptikē psukhē*] belongs to all other living creatures besides man, and is the first and most widely shared [*kai prōte kai koinotate*] faculty of the soul, in virtue of which they all have life" (415a23–26). That which is most common is the most widely shared and the first, the origin always already divided, falling apart, and as a consequence supplanted or usurped by another origin (the principle of animality). Plant-soul is a concrete expression of such division of and at the origin—the kind of primordial generosity that gives itself to all other creatures, animates them with this gift ("in virtue of which they all have life"), parts against itself, and in this parting and falling apart invites the participation of beings in the acts of living. The gift of plant-soul does not eliminate infinite differences among its recipients but allows them to surge into being, to be what they are. Because the generosity of vegetal soul is inexhaustible, at least whilst *there are* living beings at all, it is a precious non-resource, shared infinitely without being depleted, a reserve without bottom but also without depth (recall that the plant represents essential superficiality). Practices of deforestation are the ontic mistranslations of the ontological principle of infinite vegetal giving, in that such practices conflate the trees themselves, living beings that are *not* stockpiled in the planetary "factory," with infinitely renewable resources. That which is the most widely shared becomes the most deeply violated and subject to a desire for unlimited appropriation.

In spite of this, plant-soul is inappropriable, both in us and outside of us, just as the life it bestows upon "all living creatures" cannot belong to any one of them once and for all. The gift of vegetal life overwhelms the limits of our receptivity, and it is this incapacity to accept the given as a whole that instigates the (ineluctably, albeit positively, incomplete) life itself. In making this observation, we have already switched around the traditional perspective, for as we know, within the purview of metaphysics defectiveness lies on the side of the plant. According to the divisions of the human soul in *Nicomachean Ethics*, for instance, the vegetal (*phutikō*) and the most common (*koinō*) soul is depicted, in an extraordinarily privative modality, as being "without reason," *alogon* (1.13.9–10). But this depiction anticipates in advance the inversion and indeed the subversion of the power of reason: when *logos* comes across the *alogical* faculty of the soul, it finds itself faced with an absolute limit it cannot surpass, a life it cannot make its own in exercising its power of persuasion, which will potentially be effective when it appeals to the irrational (or *illogical*) part of the psyche, capable of registering the injunctions that emanate from reason. Distinct from this latter part, the vegetal, most widely shared, soul, which "does not share in the rational principle at all" (1.13.18), is absolved of all responsibility and turns into a trope of innocence,[37] unable to act otherwise than it does in growing and nourishing itself. Its acts are not immoral, but amoral; not irrational, but non-rational; motivated, in Nietzsche's words, by "the *will* to ignorance," without which "life itself would be impossible."[38] The vegetal ethics of relentless giving (of itself) passes at best for a beautiful amorality.

The consensual outcome of reasoning activity is restricted in scope when compared to the commonality (which it desperately desires to recreate) that brings living beings together in an entirely non-rational way, without a genus, by means of a "vital faculty" that "exists in all things that assimilate nourishment" (3.13.11). What is most common in all that lives establishes commonality thanks to the assimilation of nourishment, that is to say, thanks to an appropriation of the other to the same who (or that) devours it. Even so, the nutritive-physiological process does not simply map the methods of creating common grounds on the terrain of vegetal psyche. In order to avoid potential confusion, I wish to single out three versions of commonality attributed, in the history of Western philosophy, to plant-soul. With respect to each version we will assess "the common" in terms of how amenable it is to homogenization and of whether it leaves enough space for differences among the living.

i) To begin with, pseudo-Aristotle's explanation for the shared nature of vegetal life is quintessentially metaphysical in its ascription of sameness to the most fundamental stratum of the soul. The author of *De plantis* writes that "the nutritive part [of the soul] is the cause [*aitia*] of the growth of every living thing" (315b33–34). Stemming from a common cause, various individual growths—plants, animals, humans, or other living creatures—are the manifold effects of the same impulse, which is the immutable metaphysical foundation for the innumerable changes that occur in "living things." More gravely still, the static etiology of basic life attempts to tame the proliferations of the vegetal soul by confining them to the effects of a cause that can be known, and hence subjugated to the demands of the rational soul. It is as though in these lines *logos* itself makes a desperate attempt to digest and to assimilate the non-rational part of the psyche, which has staked out the impotence of reason, by explaining this part away through the metaphysical concept of causality. The common is here understood as a synonym of "sameness," as the unexceptional and the inconspicuous *par excellence*, as that which we find wherever living things are found, and as the principle of life in its utmost banality.

ii) Aristotle is more keenly aware of the difference of the vegetal soul than Nicolaus of Damascus, since his comprehension of "commonality" depends on the—admittedly Platonist—notion of participation, *methexis*, congruent with the existential idea of coexistence. "By nutritive faculty," he explains, "I mean that part of the soul [*morion tēs psukhēs*], which even the plants share [*metekhei*: participate in]" (*De anima* 413b7–8). A specific division of the soul, divided in and of itself, is shared by everything that lives; all living beings, including plants, participate in its signature activities of nourishment, growth, and procreation, though plants alone can be said to attain their proper excellence, *arētē*, in the course of this participation (observe that the virtue of plants is restricted to the three activities listed above and therefore, somewhat more abstractly, to their ability to assimilate the other *and* to become, or to engender, the other).

That this insight is inspired by Plato, for whom *methexis* is one of the privileged articulations of the Idea with its earthly instantiations, becomes obvious when Aristotle describes vegetal propagation in a way reminiscent of *The Symposium*: "For this is the most natural of all functions among living creatures . . .: *viz.*, to reproduce one's kind, an animal producing an animal, and a plant a plant, in order that they may have a share [*metekhōsin*] in the immortal and the divine in the only way they can. . . . Since, then, they cannot share in the immortal and divine by continuity of existence . . . they share in these in the only way they can, some to a greater and some to a lesser extent; what persists is not the individual itself, but something in its image [*eidei*], identical not numerically but specifically" (415a27–b9). The Platonic heritage of *De anima* is apparent not just in the usage of the word *eidos* ("idea" or "image"), which supplies the blueprint for the reproduction of various members of the same species, but especially in the aporetic logic of self-preservation, whereby every finite living entity keeps itself intact solely because it manages to replace itself with another like it, so that mortal beings would come to partake of immortality by engendering their offspring.[39] For the nutritive-reproductive part of the soul to succeed, it is not enough to assimilate the nourishing other to the nourished same; rather, the exact opposite process—the becoming-other of the same in its progeny—is decisive to a good functioning of *to threptikon*. The incorporation of the other into the same is subordinate to the othering of the same, in that the ongoing maintenance of a finite, perishable organism, dependent upon a regular intake of nutrients, does not accomplish the higher *telos* of the vegetal soul, the goal of sharing or participating in the immortal by producing another like it.

Platonic-Aristotelian "participation" implies at the same time that *to threptikon* can neither embody the divine nor become immortal, because it gains access to these qualities in a non-proprietary way, without laying an absolute claim to them. No plant coincides with the image or *eidos* it replicates (by itself or in its offspring) through its double sharing in the nutritive part of the soul and in immortality. Its approximation to its *eidos* may be confirmed only in taking stock of what it shares with the other, of what it has in common with the other as other, in whom its trace will live on. And insofar as we, humans, the others of plants or the wholly other plants, likewise participate in *to threptikon*, we complement the image/idea of a plant as a being that grows, reproduces, and shares in the divine, albeit often in a strikingly different mode.

The stress on *eidos* is a crossroads where difference and identity converge. On the one hand, plants and animals partake of immortality by way of vegetal proliferation adapted in each case to their kind of being, a proliferation that is richer, more diverse, and more differentiated than the "common cause" of life isolated in *De plantis*. On the other hand, *eidos*, taken in the sense of the immutable and transcendent Platonic Idea, casts a long shadow of sameness and identity on "the common," individuated in as many ways as there are species. The difficulty, then, lies in the ambiguity of the concept of individuality, which at the same time singularizes and generalizes the entity it comes to describe; eidetic participation is still far from an adequate expression of the non-essential mode of living-with we have inherited from plants.

iii) In keeping with twentieth-century philosophy, living is "living-with," cohabitation in a community mediated not by the immutable bonds of a common essence but by the non-essential (or better, pre-essential) difference inherent in existence. The last, and most promising, conception of commonality will correspond to this notion of community by deriving it from difference, having simultaneously dispensed with the individual as the atomic unit of analysis. Hegel and Nietzsche will be the unlikely allies in this endeavor: both will identify the quality of subindividual growth in plants that, in the words of the former thinker, are particular but, in contrast to beings endowed with the animal soul *(Seele)*, not yet individualized. Hegel goes on to argue that "in the plant the particularity is quite immediately identical with vitality in general"[40] and to refer to vegetal life as a proliferation of multiplicities, "this dispersed Spirit,"[41] which Nietzsche will rediscover by way of shattering the unity "plant" into a "millionfold growth." But what is still missing from these philosophical interventions is the thesis that instead of describing an imperfect—because open-ended—proliferation, pure multiplicity may facilitate a derivation of the common without the interference of identity, be it a shared cause or a unitary eidetic structure.

Positively understood, the dispersed life of plants is a mode of being in relation to all the others, being *qua* being-with. Dispersed in acts of living, all creatures share something of the vegetal soul and are alive in the most basic sense insofar as they neither coincide with themselves nor remain self-contained, but are infinitely divisible below the death masks of their identities. If this is so then we have a lot to learn from plants that have mastered this way of being, which is their virtue (again, in accord with the ancient meaning of *arētē*), not a vice of insufficient self-idealization and self-universalization, as Hegel would make us believe.

The shared divisibility of all living beings, first honed in the acts of the vegetal soul, pertains to the workings of the soul in general, which already in the texts of Plato and Aristotle is split, often against itself. For the psyche to live, it must receive guidance from the vegetal principle of divisibility, constantly becoming other to itself; in other words, it must be temporal through and through. But also ever since Plato, psychic principles have found their analogs in the realm of the political, even if such parallels have tended to provide a host of metaphysical justifications for a fixed distribution of power in a polity. (Plato's *Republic* is of course a dramatic case in point, given the parallel it draws between the appetitive soul and the workers, the spirited part of the psyche and the guardians, and the rational soul and philosopher-kings.) Hence, adopting Plato's psycho-politics rid of its hierarchical component, I propose the term "vegetal democracy" to designate the potential political effects of plant-soul.[42]

Both in the life of plants and in vegetal democracy the principles of inherent divisibility and participation are paramount. Inspired by the kind of sharing that marks plant-soul, which traverses all other modes of living while preserving their differences, vegetal democracy is open not only to *Homo sapiens* but to all species without exception. Like the plant-soul itself, consonant with life's hospitality, it stands for that which is most common and most inclusive, not by formally enveloping its contents but conversely by bringing into relief differences and divisions without which no "sharing," no part-icipation, no "being-with" is possible. Far from furnishing a natural or a naturalized foundation for actual and ideal democratic regimes, it is a paradigm of sharing more basic than any exchanges between "autonomous" individuals. The non-economic generosity of plant-soul, giving itself without reserve to everything and everyone that lives, transcribes vegetal democracy into an ethical politics, free of any expectations of returns from the other. Its divisibility renders irrelevant the task of reconciling particular, individual interests and the universal Good, since what happens below individual unities bears directly upon common well-being.

In sum, vegetal democracy brings together without totalizing all "growing things," that is to say, plants and the things of nature. Like plants, animals and humans too are "growing things," even if in addition to the growth of hair, nails, claws, fur, or feathers, they exhibit other kinds of growth that are experiential, intellectual, and so on. Be this as it may, participation in life—or in the slightly more restricted categories of growth and growing—is not a monolithic principle. In raising the question of the living, we must acknowledge the infinite differentiations, the "striatedness" of the field of vitality, as well as the blurring of clear demarcations between life and death in the wake of Derridian deconstruction. Spectrality (the return of a ghost who/that is neither simply alive nor dead) and survival (a simultaneous continuation *and* suspension of life) are the names Derrida bestows upon the shifting margins of life and death. Mindful of such complexity, vegetal democracy does not advocate a naïve vitalism that would insulate life and the living from death; quite to the contrary, it situates "participation in life" in an intimate relation to mortality.

We might recall here the beginning of our meditation on the meanings of plant life, namely the observation that the speculative sense of "vegetation" is paradoxical and double. "Vegetable" designates a wild and potentially untamable proliferation and at the same time veers on the side of death, in that it symbolizes immobility and torpor, not to mention the comatose condition, referred to as "persistent vegetative state," wherein life diminishes to a minimum hardly distinguishable from its opposite. Besides these semantic indications, Aristotle had already construed the mortal nature of plant-soul as the embodiment of finite life, the perishable part of the psyche that does not survive the death of the body. The life of plants is situated on the brink of death, in the zone of indeterminacy between the living and the dead. Those who share in its anarchic principle will not escape this predicament of being on the verge, suspended between life and death, the predicament common to all living beings.

If the vegetal democracy of sharing and participation is an onto-political effect of plant-soul, then it must, like this very soul, eschew the metaphysical binaries of self and other, life and death, interiority and exteriority. The plant that has no identity of its own secretly confers a plastic, malleable form upon life in its multiple instantiations and animates the grids of meaning, wherein other living beings operate. Henceforth, every consideration of post-foundational, post-metaphysical ethics and politics worthy of the name must avow the contributions of vegetal life to what contemporary approaches to the common deem so significant: the non-essentialized mode of "living-with. . . ." A reassessment of these contributions will require a further investigation into how plants quietly subvert classical philosophical hierarchies and afford us a glimpse into a lived (and growing) destruction of Western metaphysics.

2. The Body of the Plant

or, The Destruction of the Metaphysical Paradigm

Metafísica? Que metafísica têm aquelas árvores?
—Fernando Pessoa, "Há metafísica bastante . . ."[1]

Practical deconstruction of the transcendental effect is at work in the structure of the flower, as of every part, inasmuch as it appears or grows as such.
—Jacques Derrida, *Glas*

What does metaphysics have to do with plants? What can this group of heterogeneous beings, as different from one another as a stalk of wheat and an oak tree, tell us about being "as such and as a whole," let alone about resisting the core metaphysical values of presence and identity that the totality of being entails? A pessimistic response to these questions is that metaphysical violence seeking to eliminate differences—for instance, between a raspberry bush and moss, or a mayflower and a palm tree—results in a reduction of the bewildering diversity of vegetation to the conceptual unity "plant."

The plant cannot offer any resistance to metaphysics because it is one of the impoverished products of the metaphysical obsession with primordial unity, an obsession not derailed but on the contrary supported by taxonomies and the scientific systems of classification that, from the antiquity of Theophrastus and Dioscorides onward, have been complicit in the drive toward identity across hierarchically organized differences of species, genus, family, and so forth. Today the ontic manifestation of this ontologico-metaphysical consolidation of the plant is the "monocrop," such as sugar cane or corn, which increasingly displaces varied horticultures all over the world, but especially in the global South. Metaphysics and capitalist economy are in unmistakable collusion, as they militate against the dispersed multiplicities of human and non-human lives;[2] economic rationality, which currently treats plants as sources of bio-energy or biofuel, converts, concretely and on the global scale, the metaphysical principles of sameness and identity into the modes of production and reproduction of material existence. The loss of plant varieties and biodiversity is a symptom of a much more profound trend—the practical implementation of the metaphysics of the One (the Hegelian becoming actual of the rational and becoming rational of the actual) in human and non-human environments.

And yet despite the onslaught of the One, something in vegetation escapes the objectifying grasp of metaphysics and its political-economic avatars. In what follows I will argue that although in denying to vegetal life the core values of autonomy, individualization, self-identity, originality, and essentiality, traditional philosophy marginalizes plants, it also inadvertently confers on them a crucial role in the ongoing transvaluation of metaphysical value systems. It is neither necessary nor helpful to insist, as certain contemporary commentators do,[3] on the need to attribute to vegetal beings those features, like autonomy or even personhood, philosophers have traditionally considered as respect-worthy. To do so would be to render more refined the violence human thought has never ceased unleashing against these beings, for instance by forcing plants into the mold of appropriative subjectivity.

As soon as we are willing to let go of these oppressive values, we will come to realize that from the position of absolute exteriority and heteronomy, plants accomplish a *living* reversal of metaphysical values, or what Derrida terms "practical deconstruction of the transcendental effect,"[4] and thus contribute to the destabilization of hierarchical dualisms. The life of plants is—to resort to the categories Althusser used in his historical analysis of capitalism—the weakest link in the metaphysical chain, where repressed contradictions are condensed into their purest state and where worn-out justifications get so thin as to put the entire system on the verge of rupture. To demonstrate how plants silently deconstruct metaphysics and its pernicious effects will be the aim of the present chapter.

Back to the Middle: The Inversions of the Plant

Turning to the inception of Western metaphysics in Plato's thought, we witness an astounding attempt to harness the plant for the purposes of justifying the privileged theo-ontological status of the human. The highest kind of soul is housed, as Plato states in *Timaeus*, "at the top of our body," *akrō tō sōmati*, elevating us to the position "not [of] an earthly but [of] a heavenly plant—up from the earth towards our kindred in the heaven." The root of the human plant is not in the ground below its feet—since this would result in a confusion with the earthly plants that etymologically connote something driven in, if not pushed into the ground, with the feet (*plantare*)—but in the sky, in the eidetic sphere, in *topos ouranios*, the source of our humanity. "For," Plato continues, "it is by suspending our head and root [*kephalēn kai rizan*] from whence the substance of our soul first came that the divine power keeps upright our whole body" (90a). In light of the Platonic construction, our mobility is insubstantial in comparison to our invisible rootedness (indeed, our autochthony) in the realm of Ideas, the imperceptible filament that binds the top of the human body, the head, to the eidetic sphere, from which it receives its nourishment and without which the heavenly plants that we are would wither away. The soul's ground—the otherworldly soil, wherein it first sprouted—is the realm of Ideas, responsible for the sustenance and continued existence of the psyche. Only when this tie remains intact is the body itself "upright," literally and figuratively, spatially and morally, in the sense that the rational soul stays firmly in charge of the animal and vegetal desires in us.

Western metaphysics commences then with the *inversion of the earthly perspective of the plant*, a deracination of human beings uprooted from their material foundations and transplanted into the heavenly domain, and the correlative devaluation of the literal plant, mired with its roots in the darkness of the earth as much as in non-conscious existence.[5] In twentieth-century terms, honed by Sarah Kofman, one could thus say that the heavenly plant (along with its replication in the philosophical tree of knowledge) is a topsy-turvy image visible in the *camera obscura* of metaphysical ideology that demotes earthly vegetation to its own distorted reflection.

Although the morphology of the literal plant persists, notwithstanding a strong pull of idealization in the direction of ethereal reality, its spatial position and telluric attachment to the earth form the counterpoint to the metaphysical coordinates of the human. For Nietzsche as for Heidegger, it was tempting, at the end of metaphysics, to capitalize on the vegetal metaphor and to invert the Platonic inversion of material existence. Would such an overturning align our perspective with that of the plant? Not quite. Nietzschean perspectivalism, contesting the idea that there is but one objective truth, applies both to differences in perspective among human beings and to those between human and non-human living entities. Whereas from the standpoint of the human, "man" is indeed a measure of all things, for the plant, vegetal being is the standard and point of reference—"The plant is also a *measuring being*."[6] When considering "plant-thinking" and the wisdom of plants, we will revisit this insight and assess the possibility of a non-conscious access to the world from the vegetal point of view, conceived by analogy with the phenomenology of the human being-in-the-world.

For now, Nietzsche's "generalized perspectivalism," that is, perspectivalism pertaining both to individual human beings and to non-human species, complicates the project of setting the inverted metaphysical edifice aright by means of yet another inversion. Perspectival variations on truth do not leave metaphysical totality intact, but shatter it into a myriad of fragments, including the truth *of* and *for* a non-human being, such as a plant, that signifies something radically different from everything measured in human terms. The operation of a mere overturning does not suffice, both because it ignores this irreversible fragmentation and pulverization of truth and because metaphysics already anticipates its own reversals, co-opts them, and occasionally draws its energy from them. Only a non-totalizable multiplicity of perspectives, only anarchic radical pluralism comprised of the all-too-human and the other-than-human existences and "worlds" is capable of countering originary metaphysical violence, opposing the human to the plant.[7]

Heidegger too does not favor a simple overturning of metaphysics, even though his propensity to bemoan the loss of human autochthony and bucolic life might be taken as a nostalgic yearning for a plant-like existence of humanity. His 1955 "Memorial Address," celebrating the 175th birthday of the composer Conradin Kreutzer, is galvanized by positive allusions to Johann Peter Hebel's aphorism, "We are plants which—whether we like to admit it to ourselves or not—must with our roots rise out of the earth in order to bloom in the ether and to bear fruit."[8] Heidegger's interpretation of Hebel sounds in fact as though it were a direct rejoinder to Plato's grounding of the human plant in the eidetic ether: "The poet means to say: For a truly joyous and salutary human work to flourish, man must be able to mount from the depth of his home ground up into the ether. Ether here means the free air of the high heavens, the open realm of the spirit."[9] Conceptually, Heidegger's own interpretation means to say: For true talent to blossom, one must first feel at home in a familiar dwelling, domain, or culture, to which one will remain tethered even amid the highest aspirations of one's aesthetic or intellectual pursuits. Estrangement, alienation, and the sense of the uncanny, violent uprooting, diaspora, exile, or displacement would by implication hamper the flourishing of human work, as when a plant is removed from its native soil, to the extent that these phenomena interrupt the nourishing flow from the artist's native culture to her ideas.

How does this ostensibly conservative interpretation fit with Heidegger's overall philosophical program? And doesn't the plant-like image of creative genius he seems to endorse flout the strict divisions between the analytics of categories (things) and of human existence (*Dasein*) on which he insisted in *Being and Time*? The comparison of a genius to a blossom is rather deceptive, in that the very thing that makes us akin to plants increases the ontological distance between human and vegetal beings. Precisely with reference to "home ground," the alignment of the vegetal and the human perspectives crumbles: human rootedness, as a metaphor for being at home (with ourselves), is something of which the plants are incapable, because, as Heidegger points out elsewhere, unlike humans they do not dwell, do not inhabit a place, do not have any way of accessing the world.

We may of course contest this claim by arguing that plants dwell otherwise than humans and access the world differently, in a non-proprietary and non-conscious way. Still, within the framework of natural history—and especially within the scope of what we have called the "vegetal perspective," beautifully summed up in the rose's conclusion in *The Little Prince* that life must be very difficult for humans, who in the absence of roots flutter about, carried by the wind—the autochthony of human *Dasein* is already a kind of uprootedness; the flourishing of a genius and her work is incomparable to the thriving of a pine literally rooted in the earth and rising up to the ether.

Regardless of the inescapable imprecision in their alignment, the vectors of human growth in Hebel, Nietzsche, and Heidegger coincide with those of the plant, which grows from the ground up, rather than being suspended by its roots from heavens. It is then on the terms of vegetal life that we may challenge the inverted metaphysical perspective. The first stage of *Umwertung* (the Nietzschean transvaluation of old values) consists in a twofold overturning, so that everything previously esteemed as "high" is placed beneath what used to be dismissed as "low," and vice versa. It follows that the first targets of the transvaluation of values are the transcendental ideals now brought down to earth, back to their concealed roots in the sphere of immanence. It further follows that in its struggle against idealism, materialism deploys the figure and the perspective of the plant so as to de-idealize human thought and existence. The plant is the ground on which the two extreme possibilities of metaphysics fight against each other for their place under the philosophical sun. And yet as soon as they call upon a vegetal being to shore up the metaphysical edifice in any of its instantiations, the living logic of the plant quietly prevails over magnificent systems of thought, like a tree that, with its roots, puts pressure upon, raises up, and finally cracks slabs of concrete that overlay its subterranean parts. If a mere straightforward overturning of old value systems proves to be insufficient, this is because it fails to do justice to the life of plants, to human existence, and to the relation between the two.

Preempting Nietzsche and Heidegger, materialist French philosopher La Mettrie, who had argued against the idea of "plant soul," had already tried to rid the plant–human homology of transcendent overtones. In a thinly veiled attack on Plato, La Mettrie asserted, "Man is not, as some have said, a topsy-turvy tree with the brain as root, for the brain is just the joining together of the abdominal veins, which are formed first."[10] Taken to extreme, Plato's metaphors transform human beings into "topsy-turvy trees." In turn, naïve and cursory as his own actual knowledge of physiology might have been, La Mettrie recognized, with quasi-evolutionary discernment, the relative lateness of the Platonic "highest kind of soul," transposed onto "the brain," as well as the primacy of what corresponds to the appetitive or the vegetal soul in ancient Greek philosophy, transcribed into "the abdominal veins," the channels through which an organism obtains nourishment. In their development both as phenotypes and as genotypes, human beings spatially and chronologically replicate trees. All subsequent materialisms could hence be folded into this inversion of Platonism, the inversion that prioritizes (the vegetal principle of) nourishment over consciousness as the basis, the foundation, or the root of existence.

What is perhaps more surprising is that starting with the human, homologized to the plant, German idealists have undertaken still another attempt to upturn metaphysics. Goethe, Schelling, Novalis, and most notably the naturalist philosopher Lorenz Oken judged the flower, not the root, to be the highest instance of spiritual development a plant may attain, so much so that "flowers are the allegories of consciousness or the head"[11] and the "corolla is the brain of plants, that which corresponds to the light."[12] It is not so much that the flower functions as a material substratum of Spirit, a body onto which the spiritual stamp is impressed, but rather that Spirit itself submits to the flower, capitulates before it.

The blossoms of the earthly plant are the indisputable symbols of Spirit; they point beyond themselves to the ideal being of light, reason, and consciousness. Spirit turns out to be illegible, indeterminate, and mysterious in the absence of "natural," if florid, allegories and metaphors meant to clarify its workings (think back, for example, to the infamous "flowers of rhetoric"). This is why its victory is simultaneously its defeat: philosophy converts the body of the plant into symbolic space for mapping the respective places of conscious and non-conscious existences but, in the course of this conversion, it also implicitly assigns to the plant a higher position than either of the two modalities of existence it accommodates. In other words, like Spinozan substance, the plant conjoins conscious and non-conscious life, as well as idealism and materialism—the mutually complementary parts of modern philosophy.

Both La Mettrie and Oken uncritically elect to reduce the plant and its parts to structural-functional analogs to Spirit, with the end result that vegetal metaphysics initiated in "an absolute mechanistic materialism," "*un matérialisme mécaniste absolu,*"[13] becomes indistinguishable from the metaphysics of mechanistic idealism. The materialist and the idealist theses, represented by La Mettrie and Oken respectively, are two sides of the same coin: whereas the former shows that the human equivalent to the system of roots in a plant is the digestive system, the latter demonstrates that the vegetal counterpart of the brain is the flower. In each case, the "high" and the "low," enunciated in terms of value, match the respective spatial orientations and the physiological ordering of the two kinds of creatures. Thus superimposed onto various parts of the earthly plant, materialism corresponds to the roots, with their attachment to the soil, while idealism stands for the flower, the proper media of which are air and light. No longer struggling against each other, the two opposed movements of thought join one another on the common front, made of the body and the image of the plant.

The second stage in the transvaluation of values, to which Heidegger did not pay sufficient attention in his criticism of Nietzsche's "simple" overturning of Platonism, entails questioning the hierarchical arrangement of psycho-physiological elements and their roles in the living organism. Phenomenology after all teaches us that the sense of what is above and below, to the left and to the right, before and behind me, is relative to the spatial position of my body, which is not only a thing in the world but also the "ground zero," the ultimate, albeit ever-shifting, point of reference for my world. Taking this phenomenological argument to the extreme, we may demonstrate that the spatiality of all living beings—unmoored from objective determinations and emancipated from a global, disincarnated perspective that disavows its own perspectivalism—will require that a different sense of what is above and below, etc., be laboriously worked out from the standpoint of each particular life-form in question. Plants are not mere things deposited in abstract space, even if *their* lived spatiality is different from the existential spatiality of humans. Such will have been the insight of non-anthropocentric phenomenology, prompting us to consider the divergences, as well as the overlaps, between the human and vegetal relations to space as a part of the more advanced transvaluation of values in twentieth- and twenty-first-century philosophy.

The surest way to interfere with the workings of objectivist metaphysics is to strip homogenous and abstract spatiality of the absolute privilege it had enjoyed hitherto. Rather than search for a more accurate parallel to the objectively fixed head of any organism, post-metaphysical philosophy, in keeping with this ongoing transvaluation, performs a symbolic decapitation or castration of the old metaphysical values. French author Francis Ponge puts the flowers and vegetal life in general at the forefront of this effort, when he asserts that they have no head, *pas de tête*.[14] The ambiguity of the French expression, which is not as definitive as Jean-Luc Nancy's Bataillean invocation of the acephalic (or headless) discourse productive of dense non-sense,[15] should not escape our attention. Between a mere inversion and the leveling of hierarchical metaphysical oppositions, *pas de tête* can mean "no head" or it can refer to the "step of the head," given the plurivocity of the French *pas de*. . . . The expression's indeterminate, unstable meaning connotes the act of walking on one's head, feet up, or losing one's head altogether, something the author desperately desires following the example of plants: "To leave my head, to descend to the knot of being, situated . . . several centimeters below ground-level [*Quitter ma tête, descendre au noeud de l'être, situé . . . sous quelques centimètres de terreau*]."[16] This knot is of course the seed, dethroned as the originary principle, the *arkhē*, of the plant by virtue of its ownmost germination that sends offshoots both downward and upward, burrowing deeper into the earth and emerging from obscurity toward the light. Why is the ineluctable bi-directionality of growth, striving at once toward light and toward darkness, toward the open and the closed, significant for post-metaphysical philosophy? And what would it mean to write and to think in the vegetal, if not the vegetative, state, having left one's head behind or walking on one's head? What is the outcome of one's (necessarily imperfect and incomplete) approximation to the locus of vegetal being?

Ponge accentuates this seemingly banal fact of the plant's double extension when he describes the act of placing oneself in the position of vegetal being: a little below the surface and, from there, stretching up and down simultaneously.[17] One of the most compelling reasons for wishing to be in the place of the seed is, it seems to me, that germination commences in the middle, in the space of the in-between. That is to say: it begins without originating and turns the root and the flower alike into variegated extensions of the middle, in marked contrast to the idealist insistence on the spirituality of the blossom and the materialist privileging of the root. Like sentient and conscious subjects who always find themselves in the midst of something that has already begun outside the sphere of their memory and control, the plant is an elaboration on and from the midsection devoid of a clear origin. In this sense all growth is rhizomatic, if a rhizome has "neither beginning nor end, but always a middle [*milieu*] from which it grows and from which it overspills."[18] Starting from this fecund and self-proliferating station that does not obey the Aristotelian virtue of moderation, both extremities of plants are "beheaded." Plants grow from the middle, right in the middle of things, in their milieu, *in medias res*, and therefore also in the period of the *meanwhile*. The root and the flower are neither essential nor radically indispensable, having lost their metaphysical status as the spiritual culminations of vegetal being.

Let us call this phenomenon by its name: dissemination, infinitely deferring the beginning as well as the end much in advance of Derrida's writings already in Schelling's *Naturphilosophie* ("The first seeds of all organic formation are themselves already products of the formative drive").[19] The fecund middle that disseminates and decimates the origin is a spatio-temporal term; not only does it prevail over the extremes of the root and the flower, but it also defies the fiction of an absolute and self-contained past, the mythical fascination with the first beginnings. Viewed from the vegetal middle, experienced from the maddening place where it is possible to lose one's head, space and time become indistinguishable from one another. The poetic act of beheading that gives us a foretaste of the vegetal perspective does not privilege—not even negatively!—the amputated organ, put on a par with the highest as well as the lowest, the flower and the root, in the polyvocal French idiom of Francis Ponge. In sum, the organizing principle of the head loses its transcendental footing and authority.

Before attempting to assess the full impact of the transvaluation of vegetal spatiality, we must draw a sharp distinction between the middle and the center. As soon as the one is identified with the other, the head of the plant, or of any being whatsoever, is reinstated in its majestic, sovereign place, even where it does not occupy the uppermost position in the vertical configuration of the organism. In the history of philosophy the allure of both direct and inverse homologies between the plant and the human has largely depended on the upright "posture" of the plant (on what Gaston Bachelard, in the footsteps of Paul Claudel,[20] termed its "heroic uprightness [*héroïque droiture*]"),[21] which replicates human bearing in space much more faithfully than does the position of a quadruped animal.

Plato's concern with the "uprightness" of the human body had to do with its physical and moral standing, maintained so long as this body did not deteriorate to the status of the beast upon severing the head's eidetic roots. In animals that crawl or walk on all fours, the head is roughly on the same elevation as the rest of the body, but in humans it is the highest point of corporeality and so the closest to the eidetic sphere. The head's physical position confirms its authority as a center of intelligence, the sovereign decision-making organ presiding over the organism, and the radial point from which everything properly human emanates. But assuming that something else (another organ or faculty) were pinpointed as essential, the center, from which the rest would derive, would be reconstituted elsewhere in the body. (Such essentialism is what ultimately unites materialism and idealism, grafted onto the body of the plant.) Not just an isolated point, it englobes the entire organism, as in Bergson's description of the "system of nervous elements stretching between the sensory organs and the motor apparatus" and forming the "center" of animal evolution.[22] The middle, on the other hand, is often de-centered, insofar as it comprises a series of shifting and contingent "intersections" (Onfray)[23] or the "knots" (Ponge) of the here and now. It is this middle place—not a fictitious inaccessible origin—that holds the promise of growth and proliferation, dispersed from the moment of its germination. Unlike the center, it is neither gathered into a unity nor oriented in a single direction; in its sheer materiality and organicity, the plant interferes with the metaphysical fixation on the One.[24]

From the environmental perspective, the plant is itself a middle place, standing at the intersection of the physical elements: the earth and the sky, the closed and the open, darkness and light, the moisture of the soil and the dryness of crisp air. Eluding Canguilhem's definition of the living—"To live is to radiate; it is to organize the milieu from and around a center of reference, which cannot itself be referred to without losing its original meaning"[25]—vegetal beings are de-centered in their milieu, which they neither organize nor oppose. Their life does not irradiate, does not shine forth from within, and is not traceable to an essential point of reference. The vector of vegetal living moves in the opposite direction: from the outermost reaches of the elements to plants that interrelate these disparate external influences. Put in traditional philosophical terms that achieve their clearest expression in Hegel, plants are the first material mediations between the concrete universality of the earth and the purely abstract, ideal being of light,[26] though, despite the literal meaning of "photosynthesis," they do not synthesize that which they mediate. More precisely, entirely oriented toward exteriority in their diremption toward polar opposites, plants meet the elements halfway, in the middle where they serve as the media of proto-communication between diverse aspects of *phusis*. In an attempt to "combine these two worlds, the chthonic and the uranic [*réunir ces deux mondes, chtoniens et ouraniens*],"[27] they cover the earth but do not dominate or conquer it; they seek their "place in the sun" but do not usurp the places of others, notwithstanding the empirical evidence we gather from the exuberance of the jungle and everything Nietzsche has to say on the vegetal form of the will to power. The ethics of plants, *proceeding from the vegetal standpoint*, will perennially return to this middle place literally suspended *between* heavens and earth.

Commonly taken as a superfluous addition to the landscape or to the skyline, the plant makes the land and the sky what they are both in themselves and in their articulation with each other. (In the desert, void of plants, the earth and the sky are therefore disarticulated, ceasing to be themselves. Today's intensifying desertification of the earth signals the earth's, as well as the sky's, un-becoming.) If, in Heidegger's "The Origin of the Work of Art," a product of human *tekhnē*—say, the Greek temple—was capable of gathering *phusis* into a "simple manifold" of the sky above it, the cliff on which it was situated, and the violent sea underneath it, the plant is something in nature that makes nature what it is by bringing together, in a non-totalizing way, its various elements. In other words, the plant materially articulates and expresses the beings that surround it; it lets beings be and, from the middle place of growth, performs the kind of dis-closure of the world in all its interconnectedness that Heidegger attributes to human *Dasein*.[28] The tree is already a "clearing of being," even if it grows in the thickest of forests, for in its openness to the earth and the sky, to the closed and to the open simultaneously, it brings these elements into their own and puts them in touch with each other, for the first time, as that which lies below and that which stretches above.[29]

Similarly, in a peculiar mediation between the living and the dead, caressing the dead with its roots and obtaining nourishment from them, the plant makes them live again. Vegetal afterlife, facilitated by the passage, the procession of the dead (including the decomposing parts of the plants themselves), through the roots to the stem and on to the flower, is a non-mystified and material "resurrection," an opportunity for mortal remains to break free from the darkness of the earth. Thanks to the plant, fixed in place by its roots, dead plants, animals, and humans are unmoored from their "resting places"; they travel or migrate, just as in certain non-Western religions souls can find their reincarnation in plants. Unlike the crypt, supposed to keep (though it never lives up to its mission) its inhabitant in place, surrounded by inorganic matter, the grave covered by a flowerbed is always already opened, exceeding the domain of the earth and blurring the boundaries between life and death. "Flowers, culled with the dead, always for covering the coffin . . ."[30]

Vegetal Heteronomy

Gathering parts of *phusis* that by far exceed the plant in their dimensions, vegetal being maintains its radical dependence on them, so much so that its heteronomy (literally, "the law of the other," indicative of the plant's dependence on something other than itself) turns into a crucial component of vegetal anti-metaphysics.

That plants are less self-sufficient than animals is a conclusion of the author of *De plantis*, who finds it inconceivable that "the plant would be a more perfect [*teleioteron*] creature than the animal." "How could this be," asks pseudo-Aristotle, "when the animal requires no outside action in its own generation but the plant does, and needs this at certain seasons of the year? For the plant needs the sun, a suitable temperature, and even more the air. . . . The beginning of its nutrition is from the earth and the second beginning [*arkhē hetera*] of its generation is from the sun" (817a18–26).[31] In what has since become a classical theoretical move, the author of these lines blames the imperfection of the plant on its incapacity to determine itself, on its rootedness outside of itself in the external (*exōterikoū*) element on which it depends absolutely. The spatial and temporal heteronomy of plants stands in sharp contrast to the relative autonomy of animals, whom pseudo-Aristotle judges to be "more perfect" than vegetal beings as part of a valuation (and of course a corresponding devaluation) wholly imbricated with their ontological description. Perfection and autonomy are the qualities of a being that contains its origin in itself and "requires no outside action in its own generation"—that is, God or the unmoved mover—which is not at all the case for a plant.

Unwittingly, however, metaphysical discourse, marked by an emphatic abhorrence of radical dependence, guides us to the fact that the root is not one, that it is always bifurcated and thus resistant to ontological consolidation. The root, in the sense of the external cause, is split between the plant's nutritive origination from the earth and "the other origin," *arkhē hetera*—thus a certain kind of *an*archy—hidden in the generative power of the sun. Vegetal life is not autotelic; it does not contain its "cause" in itself, in contrast to the animal, which harbors the principle of its own animation. To be sure, the displacement of causality to the externality of the sun, the soil, moisture, and air still espouses the sort of "metaphysics of the element" that has been prevalent in ancient Greek philosophy already (and especially) in its pre-Socratic variations. But this displacement makes an invaluable contribution to the post-metaphysical critique of the concept of causality, insofar as it, first, disperses a unitary cause among different elements, and, second, accentuates a group of beings (i.e., the plants) that are not sovereignly self-determined and that do not assert themselves over and against their environment, wherein their causes, and hence something of themselves, lie. It is this re-conceptualization of being in terms of heteronomy, commencing from the world of vegetal beings, that offers an alternative approach to the "end of metaphysics," the unsurpassable horizon of philosophy *and* of concrete life in the twenty-first century.

The plant does not stand under the injunction, ostensibly relevant to all other types of subjectivity, to cordon itself off from its surroundings, to negate its connection to a place, so that it can fully become itself as a consequence of this oppositional stance. If vegetal being is to be at all, it must remain an integral part of the milieu wherein it grows. Its relation to the elements is not domineering: the receptivity of the flower and of the leaf is obvious in how they turn their widest surfaces to the sun,[32] while the root imbibes everything, whether nutrients or poisonous substances, it encounters in the dark recesses of the soil into which it burrows.

Of course, this hyperbolic attribution of passivity to vegetation ought to be tempered by recent findings that shed light on the way plants defend themselves from predators (for instance, by bathing the larvae of insects deposited on their leaves in toxic chemicals) and actively adapt to changes in their environment. It would be more accurate to describe plants as neither passive nor active, seeing that these behavioral attitudes are merely human projections onto the world around us. Western philosophers of subjectivity, nonetheless, used to associate vegetal life with a purely passive comportment and regarded it as deficient, since it did not operate in the space of freedom to decide on the course of action or, for that matter, to act. In the context of the post-metaphysical rethinking of ethics in the writings of Levinas and Derrida, such radical passivity in excess of the opposition between the active and the passive, such exposure to the other, typical of plants, which is affirmed well in advance of *our* conscious ability to utter a decisive "yes" or "no," denotes the ethical mode of subjective being. Opening themselves up to the other, ethical subjects prompt the plant in them to flourish. While plant existence is ethical, post-metaphysical ethics is vegetal.

Nowhere is the tacit philosophical disagreement on the subject of receptivity as evident as in the divergence of the Levinasian ethics of alterity from the Hegelian emphasis on self-relatedness. The two millennia separating pseudo-Aristotle and Hegel did not see any substantial changes in philosophical approaches to plants, proving once again that the ontology of plants has always been elucidated in the shadow of metaphysics, with its adherence to the non-temporal thinking of being.

As though echoing the ancients, Hegel deplores the non-oppositionality of plants and their absolute dependence on external conditions, the determination of their movement by "light, heat, and air."[33] Although he falls short of stating that plants are devoid of selfhood (*Selbstischkeit*), the German thinker terms the vegetal self "negative," because "the plant is not yet self-related," because, that is, the "outer physical self of the plant is light towards which it strives in the same way that man seeks man."[34] In other words, the plant is unable to lay hold of a self, to come back to itself, and to become conscious, because it is *objectively split* between the growing thing *simpliciter* and the objective conditions of possibility for its existence that stand for its true "outer physical self."

The Hegelian plant lacks a sense of self precisely because of its inability to relate to itself other than by striving to this outer self in a way that is phenomenally obvious, extended, and material. Nonetheless, growth does not bridge the objective gap between the physicality of the plant and external elements, let alone the subjective divide between the plant's negative self and the world, wherein, outside itself, it seeks fulfillment. The anthropocentric take on the striving of the plant to its other ("in the same way that man seeks man") is at the same time a limit to the similarity between the two kinds of being, supposing that humans indeed find their fulfillment and relational selfhood in other human beings and ultimately in themselves. This difference highlights the divergence between two models of heteronomy: the radical and incorrigible dependence of plants on their others that are not at all akin to them, and the relative dependence of human beings on other humans, leading up to the emergence of relational autonomy.

At the basis of the Hegelian example we find the sort of heliotropism that, as Derrida explained, has been plaguing philosophy at least since Plato's analogy of the sun and the Good.[35] We seek other humans (and more abstractly, knowledge itself), just as the plant unconsciously looks for the light of the sun. Positioned between plants and humans, animals are also included in this metaphysical quest, if, as Hegel, Bergson, and others have claimed, their concrete will corresponds to the vegetal striving toward the light. In the words of Bergson, the parallel runs as follows: "We doubt whether nervous elements, however rudimentary, will ever be found in the plant. What corresponds in it to the directing will of the animal is, we believe, the direction in which it bends the energy of the solar radiation."[36] The will and the bending of "the energy of the solar radiation" betoken the intentional acts of different kinds of life, acts related to the particular purposes of the living beings themselves.

But even this formal analogy falters: while the will presumably emanates from the interiority of subjective being, the plant's processing of solar energy originates with the exterior shining of light; whereas the plant, in its proto-willing, does not furnish itself with a delimited object, animals and humans direct themselves to a circumscribed piece of reality meant to satisfy their needs. This non-objectification of the real is perhaps the crux of the non-domineering relation of vegetal beings to their environment and the obscure echo of contemporary philosophy's longing for a view of the world unfiltered through the modern categories "subject" and "object." Bergson's own frequent insistence on the errors and illusions for which the objective delimitation of the real is responsible could be productively combined with the plants' mode of being immune to the pitfalls of such delimitation. Learning from plants is in the first instance unlearning the objectifying approach to the world.

The lesson Hegel draws from the plants' quasi-religious striving toward, if not the worship of, the light of the sun immanently undoes the logic of self-centered subjectivity. We have already glimpsed certain moments of this undoing in the discussion of the vegetal soul. As opposed to human subjects, who attain their subjecthood thanks to a return to themselves across the terrain of otherness they have traversed, the formation of the vegetal self proceeds in the absence of self-reflection or fully developed self-feeling,[37] in and as a unidirectional, infinite movement toward its other, namely light. But what Hegel, in his verdict, denounces as the "bad infinity" of plant life is the very ethical infinity that resists the logic of totalization in Levinas. The infinite relation to the other without return to oneself is the touchstone of Levinasian ethics and of the associated ethical approaches that advocate the substitution of the appropriative model of subjectivity with the receptive orientation to the other. The plant embodies, *mutatis mutandis*, this approach to alterity, in that it tends, with every fiber of its vegetal being, toward an exteriority it does not dominate. Its heteronomy is symbolic of Levinas's quasi-phenomenological description of the subjectivation of the I in an ethical relation to the other that/who is unreachable and cannot be appropriated by the I. Vegetal heteronomy, therefore, holds the blueprint for the formation of subjectivity at the current stage of post-metaphysical thinking, with its emphasis on the constitutive role of the relation to the other.

The conceptual correlation of plant ontology and Levinasian ethics should be mindful of its own limitations, especially when it comes to issues of time and space. While the plant is an integral part of its surroundings, in *Totality and Infinity* the ethical subject sets itself apart from the elemental world in a separation meant to establish its psychic interiority, whence the movement toward the other will commence. Despite the prevalence of the language of spatiality here (interiority/exteriority, separation, etc.), the core of ethical subjectivity involves a temporal, not a spatial, dimension of existence. For Levinas, space, rather than time, is the domain of sameness, a relentless contiguity where differences are superficial and merely quantitative. But isn't spatiality the exclusive province of vegetal life? In the absence of self-relation, or what Kant calls "auto-affection," does it really give rise to a temporal order through the formation of negative selfhood? It would be prudent to bracket, for the time being, the presupposition that the plant is an entirely spatial, extended being, excluded from the order of temporality. The time of plants is a sensitive topic requiring a patient analysis and elaboration.[38]

Given that the plant is not separate from its environment, both Hegel and Levinas will find questionable the suggestion that it could relate to alterity at all; at best, Hegel will admit that "[the plant's] other . . . is not individual, but what is elementally inorganic" and therefore what is other to life itself.[39] The non-individuation of the vegetal self is symmetrically reflected in the non-singularization of its "elementally inorganic" other. The exigencies of individuation that do not apply to the plant form a metaphysical foundation for relationality and for ethics. But if the other of vegetal life is the inorganic world *in toto*, then how to think about the relation (or, at times, lack thereof) between plants and animals, let alone between plants and human beings?

We are not, strictly speaking, the others of plants, since we obviously do not fall under the category of the "elementally inorganic," nor are we the same as they are, though we do participate in many of the processes defining vegetal soul. When humans interfere with the conditions of vegetal growth—for instance, by altering the temperature of a hothouse—they come to mediate the unidirectional relation of plants to their other. Such interference may also be indirect and perhaps unintentional, as in the desertification of vast areas of the globe, partly attributable to human activity and in any event detrimental for vegetal life. Or it may be barely noticeable when we merely contemplate wildflowers during a walk in the woods. There is then no fixed mold or model for the relation of human and vegetal beings, since the degrees to which the former mediate between plants and their "proper" inorganic other vary. To the extent that we practically engage with vegetal beings, we interpose ourselves in the place of what is other to them, the place that does not inherently belong to us. Human usurpation not only of our place in the sun but also of the very place of the sun vis-à-vis plants (not to mention all the other material conditions of possibility for growth) is increasingly the source of our metaphysical domination over them today.

The insistence on separation in Levinas is itself a vestige of the metaphysical tradition, which is unwilling to let go of the presupposition that, phenomenologically, experience starts with a free and autonomous subjectivity oblivious to its heteronomous provenance. Levinas wants above all to demonstrate that extreme egoism is unsustainable and how even the most heedless retreat into the I thrusts this I toward the other. But, *concesso non dato*, shouldn't an ethically receptive subject forgo, as its incipient moment, the very principle of appropriation and the view of subjectivity as a hidden repository or a storehouse of experience, if it is to be genuinely generous? Vegetal life is capable of this because it is bereft of interiority, because—with a few exceptions—it does not necessitate the consumption of other living beings, and because, as pruning exemplifies, the more the plant loses, the more it grows. Proliferating from pure loss, plants offer themselves with unconditional generosity. Silently they extend themselves in space, exposing their vegetal bodies in utter vulnerability to being chopped off or plucked, harvested or trimmed, broken by a hurricane or burnt by the sun. Ethical humanism will interpret such selflessness as an unattainable ideal only there where the critique of metaphysics still does not disturb the ideal of possessive subjectivity. But as soon as ethics sheds its humanist camouflage, the human subject will join plant life in a self-expropriating journey toward the other.

The Language of Plants and Essential Superficiality: An Approach to Vegetal Being

The other remnants of metaphysics in Levinas's *oeuvre* revolve around his prioritization of speech over writing as a responsible and ethical mode of addressing the other. Along with the modulations of the breath that produce it, speech is offered to the other, such that the speakers do not have a chance to hide, to erase themselves from the scene, to dissimulate themselves behind the words they utter. Behind the indispensability of speech for ethics is the voice—or more precisely, the arch-phenomenon of hearing-oneself-speak—coded as the ideal medium of subjectivity that coincides with itself in an auto-affective key following a philosophical genealogy traceable from Husserl back to Hegel and Aristotle.

The plants, on the other hand, are voiceless and consequently cannot address the other, let alone auto-affectively coincide with themselves. Despite its obviousness, Ponge's nearly phenomenological description in "Fauna and Flora" is philosophically noteworthy: "they [=the plants] have no voice [*ils n'ont pas de voix*]."[40] They can certainly make sounds in conjunction with the elements, as in the case of wind passing through the reed or a bamboo grove, and they can send biochemical signals in response to altering environmental conditions, but the silence of vegetation is unbreakable and absolute, because, deprived of the possibility of speaking, it does not keep anything back, does not conceal anything. The muteness of plants puts up insurmountable resistance to the mechanisms of subjective self-idealization that permit the subject to be present before itself in the closest proximity of hearing-itself-speak. The plant's non-coincidence with itself is an effect of its silence.

Vegetal life expresses itself otherwise, without resorting to vocalization. Aside from communicating their distress when predators are detected in the vicinity by releasing airborne (or in some cases belowground) chemicals,[41] plants, like all living beings, articulate themselves spatially; in a body language free from gestures, "they can express themselves only by their postures [*ils ne s'expriment que par leurs poses*]."[42] In using the word "language" to describe vegetal self-expression in all its spatialized materiality, I am not opting for a metaphor. What I propose instead is that contemporary philosophy include plants in the tradition of treating language neither as a means of communication nor as something exclusively human. The language of plants belongs in the hypermaterialist tradition that (in Heidegger's "totality-of-significations" and in Walter Benjamin's "language of things" or "the language as such") is alive to the spatial relations and articulations between beings, animate or inanimate. Plant-thinking, in turn, cannot but rely on material signification that bypasses conscious intentionality and coincides with the very phenomenality—the modes of appearance—of vegetal life.

If the postures of plants are meaningful in the strict sense of expressivity within the context of Ponge's poetic thought, then it is possible to appeal to their embodied, material, and finite sense as the inverse of the ideality of meaning endorsed most bluntly in Husserlian phenomenology. Recall that it is with reference to the destructibility of an actual tree, as opposed to the noematic perceived tree, that Husserl endeavors to formulate the metaphysical sense of sense in *Ideas I*: "The *tree simpliciter*, the physical thing belonging to Nature, is nothing less than this *perceived tree as perceived* which, as perceptual sense, inseparably belongs to perception. The tree *simpliciter* can burn up, be resolved into its chemical elements, etc. But the sense—the sense of this perception, something belonging necessarily to its essence—cannot burn up; it has no chemical elements, no forces, no real properties."[43] The founder of phenomenology takes it for granted that the tree *simpliciter* is as such meaningless and that, in its pure ideality, meaning is metaphysically safe and sound, insulated from empirical accidents and phenomenological reductions, devastating fires and calculated "bracketings." The ideality of sense that outlives the destruction of its referent is a corollary to "pure consciousness" that survives the hypothetical annihilation of the world.

But what if these assumptions are unwarranted? What about the materiality of the tree's sense, indistinguishable from its very being? Can it "burn up, be resolved into its chemical elements, etc."? An affirmative answer to these questions will envisage the sort of non-metaphysical meaning that will be subject to destruction along with its "bearer." It will assert that vegetal life is laden with meaning in all its spatiality, materiality, and finitude. And it will imply that the loss of every tree—to deforestation or to other causes—equals the passing away of meaning bound up with the particular spatial extension of that very tree, to which we can no longer afford to be transcendentally indifferent. Just as in the radical empiricism of Levinas and Derrida the death of each singular human being is nothing less than "the end of the world" (phenomenologically and ethically speaking), so the uprooting of every tree signals the obliteration of the meaning that it *is* in the extended materiality of its posture.[44]

The Husserlian tree *simpliciter* means *too little* (indeed, nothing at all) before the act of sense-bestowal that at the same time holds for it the quasi-dialectical and crypto-religious promise of resurrection and immortality as it turns into the perceived, the remembered, or the signified tree, immune to empirical calamities. This is why a meaningless material plant must be sacrificed to the ideal signification, through which it will gain eternal life. The destruction of the actual flora that, for Hegel as for Husserl, exemplifies meaninglessness *par excellence* does not detract from the transcendentally insulated meaning; in fact, it only reinforces the pure metaphysical ideal unperturbed by changes in empirical reality. But the question—and this is no idle speculation—is what will happen to the noema "tree," after the last tree *simpliciter* is destroyed? To what extent can the signified tree persist in the absence of an actual tree growing in my backyard, in the unique Costa Rican cloud forest, or anywhere else in the world? Even if we erroneously assume that the plant itself is meaningless, the ideal meaning and its noematic permutations still require at least the possibility of a perceptual approach to the plant as the phenomenological basis for (and the deepest stratum of) all other noematic operations. The virtualization and idealization of vegetal meaning goes hand-in-hand with the actual destruction of plant life; something of meaning always burns up along with the plant itself, even one as meaningless as the Husserlian tree *simpliciter*.

From the standpoint of classical metaphysics, a human being too may descend to the meaningless of the plant, most of all when producing *too much* sense, in excess of the confines of formal logic: "For such a man [refusing to reason and not respecting the principle of non-contradiction], as such, is seen already to be no better than a mere vegetable [*homoiōs gar phutōi*]" (*Metaphysics* 1006a12–15). It is irrelevant whether the targets of Aristotle's attack succumb to the impulses of the vegetal part of their souls. What matters is that the refusal to reason is wedded to the "possibility of the same thing being and not being," the possibility of snubbing the principle of non-contradiction and declining to master or to eliminate the equivocality and dissemination of meaning. Be this as it may, Aristotle's metaphysical ontology here exhibits its dynamism, insofar as the specific type of each being is concerned. What makes humans human is not their physical shape but *logos* in its multiple senses, while the loss of this distinguishing element renders them ontologically akin to non-human beings.

Remarkably, Aristotle skips over the level of animality in describing the downfall of someone who does not respect the principle of non-contradiction and who becomes "no better than a mere vegetable." Why? Because, following the Aristotelian train of thought, in the absence of the reasoning capacity proper to them, humans stand below even animals, by nature devoid of this capacity. The descent down to plants is thus meant to bring to the fore the calamitous ontological consequences of the betrayal of formal logic, the betrayal that is a synecdoche for the loss of reason and of sensibility permeated by reason in human beings.

Derrida's take on this passage from Aristotle's *Metaphysics* emphasizes how metaphysical thought, in vigilantly warding off the slightest hints of metaphoricity, gets carried away when it comes to a subtly denaturalized metaphor of the plant: "And such a metaphorical vegetable (*phutos*) no longer belongs completely to *phusis* to the extent that it is presented, in truth, *by mimesis*, *logos*, and the voice of man."[45] (Decades later Derrida will apply the same argument to the exemption of the metaphor of the beast from the order of nature, given that the attribute of *bêtise*—stupidity, folly, foolishness: from the French *la bête*, the beast—does not describe the animal, only the human.)[46] A dangerous metaphoricity has already percolated into the comparison—undertaken in all seriousness within the confines of a metaphysical discourse wholly committed to formal logic—of an unreasonable human being to a vegetable, in Giorgio Agamben's words, "obscurely and absolutely separated from *logos*."[47] The metaphor of the plant, wielded as a weapon against metaphorical thinking, announces the self-undermining and the internal collapse of Aristotle's dream of univocal meaning. Nevertheless, to be fair to Aristotle, the plant into which an illogical human being turns is not a metaphor but the ontological consequence of subtracting *logos* from a thinking-living being. Rather than a rhetorical figure of speech that survived its insertion into metaphysical discourse, the reference to the vegetable status of the one who disrespects *logos* is itself a logical upshot of negating that which is most human in the human.

As Derrida remarks, the Aristotelian vegetable-human is a plant that no longer belongs squarely to *phusis*, and, we might add, the Husserlian tree *simpliciter* occupies the same ambiguous place in-between. An example of "a physical thing belonging to Nature," it is also a counterexample to the phenomenological argument, wherein it makes its appearance: as such, the tree *simpliciter* is a semantic unit (featuring a Latin locution, posited in and manipulated by a philosophical text, etc.) despite its pretense of being a referent exterior to the discourse that names it. Its "simplicity," or detachment from the order of signification, is never simple enough and never complete because phenomenological discourse overwrites and overrules it, investing it with the meaning of sheer meaninglessness.

In the wake of Husserl's failure to think the tree, we must look for a simpler version of "simplicity," which is a clue in the search for the language/meaning of plants and their corresponding ontology. Although Portuguese poet Fernando Pessoa still attributes a kind of metaphysics to vegetal life, his is a metaphysics that is no longer anthropocentric, that does not revolve around symbolic meaning, but rather revels in the "clear simplicity / And health in the existence / Of the trees and of the plants [*clara simplicidade / E saúde em existir / Das árvores e das plantas*]."[48] Alberto Caeiro, the pastoral literary persona, or heteronym, of Pessoa, exhibits extreme sensitivity to the otherness of vegetal meaning in his refusal to consider the absence of conscious thought as the sign of absolute meaninglessness. The "clear simplicity," *clara simplicidade*, of plants is simple, above all from their own perspective (verging on a lack thereof) of a healthy existence that does not problematize itself and therefore does not become an issue for itself. A language that is equally simple—the language of Caeiro in the citation as well as in the rest of his poetry—imitates the simplicity of vegetal existence, from which it is, however, separated by the abyss of a symbolic expression that yearns for its own annihilation and for the collapse of the partitions between it and the simple metaphysics of plants. This yearning is of course foredoomed, because it spells out the end of poetry and a descent into silence, which mimics but still falls far short of the absolute muteness of plants.

As though anticipating the impossible desire of Francis Ponge, who wishes to lose his head, emulating the headless plant, Pessoa praises vegetal metaphysics: "Metaphysics? What metaphysics do those trees have? / To be green and lush and to have branches / And to bear fruit at the right time . . . / But what better metaphysics than theirs, / The metaphysics of not knowing for what they live / And of not knowing that they do not know it?"[49] Much more is at stake in the "metaphysics of trees" than the conventional ascription of innocence to vegetal life; what Caeiro hints at is the non-privative dimension of a thought-free way of being and a corresponding indication that thinking itself is a defect or a disease, an anomaly or a gap, within the order of existence. Implicitly concurring with both Nietzsche, for whom empty contemplation is an illness, and Heidegger, who supports the view that the theoretical attitude is a sickly product of a lapse in the practical comportment of the ready-to-hand, the author of these lines doubly affirms the disconnect between life and formal knowledge, between existence and the brooding about its meaning or purpose ("The metaphysics of not knowing for what they live / and of not knowing that they do not know it"). The spatial and practical vegetal meaning bereft of consciousness and self-consciousness completes the de-centering of metaphysics in plant life and proudly parades the lack of inner essence that would be hidden behind the surface level of its phenomenal appearances. What we can learn from the plants' "clear simplicity" is an inherently superficial mode of living, which we humans are rarely satisfied with, even though we carry on such existence during a significant portion of our lives.

The Pessoan ideal of subjective superficiality is only one half of the ontological simplicity of vegetal being. The other half may be termed "objective superficiality," expressing the sheer phenomenality and exteriority of plants (to themselves). Metaphysically speaking, vegetal life is objectively superficial because it does not boast a deep essence, because, that is, a plant may cast off virtually any of its parts without being fatally affected by this loss. The author of *De plantis* was the first to remark upon this puzzling behavior of various parts of plants, such as the leaves and the fruit, which "often fall off from them even without being cut off" (318b10–15). It is nothing out of the ordinary for the plant to fall apart, to fall off with or from itself, without compromising its existence; its provisional assemblage does not overlay a deeper unity, does not hide a more profound—metaphysical—source of organismic life. Anachronistically put, the objective superficiality of plants is indicative of their post-metaphysical, self-deconstructive ontology.

Unlike Theophrastus, who at the very outset of his *Enquiry* praises the complexity of plants that seem to have no essential parts, pseudo-Aristotle is uncomfortable with this conclusion. The latter thinker refrains from defining the entire vegetal being as essence-free, instead highlighting the parallel between these "detachable" parts, shed without damaging the rest of the living creature, and the superfluous nails and hair of human beings. He brings his thinking back into the familiar metaphysical fold (and plunges it into the dimension of depth) as soon as he contemplates the role of the root as "the source of life [*aitian zōes*]," and of the stem, in its erection out of the ground, as "comparable to the stature of man" (319a23–25). Periodically shedding its leaves does not present a danger to the continued life of the plant, which actually survives the harsh seasons thanks to this sacrifice. But detach it from its vital source (or "cause") and it will immediately perish—or so assumes the early vitalist essentialism, inaugurating the view that both the plant and the animal are organisms, or living totalities, whose organic parts are subordinate to the demands of the whole. Indeed, the invisible root, groping toward the physical depths of the earth, will itself become a symbol of metaphysics, interchangeable with the idea of the primordial source of epiphenomena that comprise tangible reality.

But is the root any different from the other essentially superficial parts of plants? It is an achievement of Derridian deconstruction to have revealed that the detachable, "prosthetic," and ostensibly superfluous supplement is the disavowed source of that which it supplements. In the plant, the leaf is the very embodiment of supplementarity, because it is something superadded onto the trunk and the branches, more often than not on a temporary basis. Goethe's *The Metamorphosis of Plants* plays out the logic of the deconstructive supplement *avant la lettre* with respect to the status of the leaf in the development of plants. According to Goethe, metamorphosis, change of form, the process of becoming-other, is not just one among many features of vegetal life; it *is* this life itself. In this influential botanical monograph he deduces the primacy of change over the stability and identity of the plant from the permutations of the leaf, whose thickening contraction yields a seed, whose refinement turns it into a petal, and whose "greatest expansion" accounts for the appearance of a fruit.[50] The depth of the root, the fruitfulness of the seed, the thickness and overwhelming size of a tree trunk are all explicable with reference to the rhythmic vacillations of the leaf, which successively experiences phases of expansion and contraction. They are all variations on a superficial supplement that constantly becomes other to itself.

Two implications for the ontology of vegetation follow. First, like human corporeality, the plant's body is all skin, a mere surface, sometimes thin to the point of transparency, sometimes thick and dense, as though in commemoration of the inorganic nature to which it stays relatively close.[51] Second, the mystical aura of the seed taken to be an originary principle—suffice it to mention, in this respect, the pre-Socratics' fascination with *spermata* and Ovid's haunting description of originary chaos as full of "warring seeds [*discordia semina*] of ill-matched elements"[52]—dissipates in the aftermath of the overturning of causal relations (the effects become the causes of the cause) and of priorities (the first becomes second, and the second, first) in the logic of supplementarity. That which is the most superficial takes the place of the most fundamental. The leaf usurps the originary status of the seed.

When Goethe resolutely argues for "the fruitfulness hidden in a leaf,"[53] he both embraces the absolute superficiality of vegetal being and adds a new twist to the cryptic statement of Saint Thomas Aquinas, "Life is hidden in plants." The mystery of this life does not lie in the deep recesses of the seed and of the earth, for it resides right on the surface, that which is given to sight and turned toward the light, the trope of being-exposed, the leaf. But the anti-metaphysical bend of Goethe's text is eclipsed by a lamentable imposition of identity onto the plant, whose differences in form are traceable to the self-same substratum underwriting them: "The process by which *one and the same organ* appears in a variety of forms has been called *the metamorphosis of plants*."[54] And, again: "Earlier I tried to make as clear as possible that the various plant parts developed in sequence are intrinsically identical despite their manifold differences in outer form."[55] A series of crystal-clear distinctions between the inner and the outer, the one and the many, the unapparent identical core and the appearance of "manifold differences" recovers the organizing set of metaphysical dichotomies that suffocate and philosophically misappropriate vegetal life. Goethe's metaphysical zeal thwarts even his careful displacement of the origin, deducible from the relegation of fruitfulness to the leaf.

Having developed, in theory, the primal form of the plant, Goethe embarked on a search for the archetypal plant, *Urpflanz*, which would supply the empirical proof for his theory. Thus, in a letter from Naples dated "May 1787," he confided in Herder: "The primal plant is going to be the strangest creature in the world, which nature itself will envy me."[56] It is going to be, more concretely and in an eerie anticipation of genetic manipulation targeting vegetal DNA structure, an actually existing generic blueprint of plant-being, from which the as-yet non-existent plant varieties could be derived. Or, philosophically speaking, it is going to be not a creature at all but an ideal unit: *the* plant, *the* leaf, and therefore a concept generative of concrete beings necessarily defective in comparison to it.[57]

The metaphysical idea that vegetal difference is inessential migrated from Goethe's theory of metamorphosis to Hegel's dialectical enunciation of plant nature. Hegel astutely recognized that, although it is an organic being, the plant is not an organism, given that in it "the *difference* of the *organic parts* is only a superficial *metamorphosis* and one part can easily assume the function of the other. . . . In the plant, therefore, the members are particular only in relation to each other, not to the whole; the members themselves are in turn wholes, as in the dead organism where in sedimentary strata they are also external to one another."[58] For Hegel, the difference of plant parts is no difference, since it is predicated on "a superficial *metamorphosis*" overlaying the undifferentiated substratum of nascent organic life still in a tight grip of the inorganic mineral world. But neither does the language of sameness befit vegetal life, seeing that the plant falls short of positing its self-identity in a mediated relation to itself as other. At the very least, the inapplicability of either of the two terms should have given the philosopher pause and should have led to the conclusion that metaphysical umbrella categories do not cover this kind of life, lived on the hither side of the dialectics of the same and the other, identity and non-identity, individuality and anonymous existence.[59] It would be fair to say that the plant immobilizes the dialectical back-and-forth from within, that its movement is too agile to be reflected on the radar screens of dialectics, and that it necessitates an alternative approach, ethically and epistemologically in tune with its absolute (i.e., non-relative and non-relational) difference.

On the dialectical terrain, however, the externality of parts in relation to the whole and to each other engrains death itself into vegetal life, as Hegel shows. The plant, considered through a Hegelian lens, is a novice in the sphere of the living, which the German thinker identifies with the organicity of a self-proliferating totality. There is nothing new either in conjuring images of death to describe the vivacity of plants, or in situating vegetal beings in ontological proximity to the world of minerals. What is of interest to us is rather the imputation of a sedimented superficiality ("in sedimentary strata") to the being of plants that bears a close resemblance to the "thousand plateaus," where, despite the superimposition and overlapping of multiple levels of sense and non-sense, the dimension of depth is absent. These countless superficies of vegetal being form the scene where the life of plants unfolds outside all organismic mediations that, contrary to what Hegel believes, act to stifle that which is living. To live is to be superficial and dis-organized: to exist outside the totality of an organism: to be a plant.

If the plant is not an organism consisting of interdependent organs, we should avoid conceiving it as a totality or as a differentiated whole. Its parts likewise transcend the distinction between "part" and "whole"; in their externality to one another, they are both members of a plant and independent entities in their own right. Unbound from the logic of the totality, they constitute a provisional unity of multiplicities ("The plant," in an apt expression of nineteenth-century French botanist Brisseau-Mirbel, "is . . . a collective being"), a loose community not interlaced with the ironclad ties of an inner essence.[60] To sum up: the plant's parts escape the grasp of the Hegelian Notion and the nets of conceptuality.

The assertion that vegetal being is essentially superficial presupposes the idea that the plant, whose forms and functions are fluid, is not an organism but what Deleuze and Guattari term a "body without organs," a mode of dis-organization, "a pure multiplicity of immanence."[61] It is all the more astonishing then that the authors of *A Thousand Plateaus* single out a particular kind of plant (the tree) as the exemplar of a hierarchical arrangement of multiplicities and of the differences between products and reproductions (tracings), between the originary and the derivative elements: "The tree articulates and hierarchizes tracings; tracings are like the leaves of a tree."[62] Deleuze and Guattari forget that the leaf is not an organ of a larger whole and that it is far from being a derivation from the original stem–root structure. In and of itself, it is an infinitely iterable and radically egalitarian building block of the tree, for it is at once the source, the product, and the minute reproduction of vegetal being, from which it may at any time fall away. Wreaking havoc in the differential valuations of copies *versus* originals and enacting a veritable anarchy, the plant's "body without organs" does not evince hierarchical organization. It maintains conceptual horizontality even in the tree's spatial verticality.

Furthermore, the suggestion that the plant is "a collective being" implies that its body is a non-totalizing assemblage of multiplicities, an inherently political space of conviviality. For the concept of the *body politic* to be germane to vegetal democracy, this body needs to be sharply distinguished from the organism whose parts—the organs—are subservient to the demands of the whole. Post-metaphysical vegetal thought ought to resist a double projection: on the one hand, of the animal and human constitutions onto plants said to possess parts that are homologous to the organs of other living creatures; and, on the other, of a contrived organicity of nature, conceived as a living whole, onto the *socium*. Where philosophers do not respect the first principle of resistance, they ascribe redundancies and superfluities to plants that are not differentiated into organs; where they slight the second principle, a totalitarian socio-political system emerges on the basis of the idealized organic sphere. A thorough de-naturalization of society and politics is possible only after a painstaking denaturalization of nature itself (to which vegetal life may contribute a great deal) and a meticulous cataloguing of its existential features that, in *Being and Time*, Heidegger reserved for the human *Dasein*. In their mutual, albeit uneven, interpenetration, the existential facets of plant life and the vegetal heritage of human existence do not provide decisive evidence for "the uniform variety of nature."[63] They instead shake up the metaphysical distinction between sameness and otherness, so that this distinction's explanatory power no longer extends either to the life of plants or to the relation between human and nonhuman "natures."

After this brief methodological detour, we are ready to return to the Goethe–Hegel nexus. The most significant disagreement between the two German authors, having to do with the philosophy of plants, pertains to the status of sexual difference (and more generally, organic differentiation) in vegetal life. Along with (1) the subjective and objective superficiality of plants, (2) the originary supplementarity of the leaf, and (3) the non-organismic arrangement of the "parts" of plants, (4) the proliferation of sexual differences in vegetation will be crucial for our understanding of post-metaphysical vegetal being. Each aspect of the ontology detailed here offers resistance to the totalizing tendencies of metaphysics, as well as a concrete refutation of the metaphysical yearning for the pure origins and first causes of life. The superficiality of vegetal being is particularly effective in dispelling this last myth, concerned with pure origins, to the extent that its unique sexuality disseminates both the fictitious unity of the first cause and the faulty notion of linear causality. More importantly, a non-metaphysical reconstruction of plant ontology will liberate sexual difference from its confinement to a binary opposition of the two sexes and breathe new life into the phenomena of dispersed, perverse, and non-productive sexualities. Vegetal sexuality and the logic of supplementarity will henceforth reinforce each other.

In Goethe's theory, the metamorphosis of plants is a teleological development, in the course of which the leaf undergoes a gradual refinement and even "spiritualization" in its transformation into the flower, a garland surrounding its sexual organs. "By changing one form into another," notes Goethe, "it [the plant] ascends—as on a spiritual ladder—to the pinnacle of nature: propagation through two genders."[64] The *telos* of the leaf, in its literal and metaphorical journey out of the coarseness of the seed and the darkness of the soil toward the vast airy expanse, toward the light, and toward the objectification of the luminous in the colorful fragility of the flower, is individuation and sexual difference seen as the basis for the ontological difference between spiritless matter and actualized spirit. The plant faithfully corresponds to the auto-teleology of nature itself, the "pinnacle" of which is self-reproduction by means of "propagation through two genders." On this account, the plant develops sexual *organs*, while the iterations of the same—the leaf—produce qualitatively different, organically differentiated, outcomes.

It is this double supposition that Hegel dismantles, arguing that the plant is unable to muster enough individuality so as to oppose itself to a specimen of a different sex: "The different individuals cannot therefore be regarded as of different sexes because they have not been completely imbued with the *principle* of their opposition."[65] Although sexual difference surfaces for the first time in a plant, it thereby signals a dialectical transition to what is not a plant—to the animal, completely imbued with sexuality, which is inseparable from its entire embodied being. Of course, Hegel does not deny that pollination is a sexual mode of reproduction, but he treats it as inessential and redundant within the overall framework of the existence of plants: the entire vegetal "genus-process" is "on the whole, superfluous since the process of formation and assimilation is already reproduction as production of fresh individuals,"[66] so much so that the "seed which is produced in a fruit is a superfluity."[67] What stood for the pinnacle of the plant's spiritual development in Goethe turns out to be a superfluous appendage in Hegel's philosophy. Notwithstanding the mediation of dialectical self-understanding through vegetal metaphors (first and foremost, the interplay between the deep essence and the appearances as a relation between the kernel and the shell), the seed of the actual plant comes to symbolize luxurious excess, ready to be discarded by dialectical machinery at any moment. The most essential is disclosed as "superfluous."

It is easy to forgive Hegel for his ignorance of the much-later discoveries that the sexuality of plants is so complex that it is regulated by hormones—for instance, soy beans contain large quantities of phytoestrogens, similar to human estrogen—or that the introduction of mammalian sex hormones into plants induces flowering and affects the ratio of female to male flowers.[68] What is unforgivable, however, is that his approach to vegetal sexuality condenses in itself, as though in a philosophical microcosm, the metaphysical mishandling of plants. The absence of individuality, inner differentiation, and oppositionality in vegetal being boils down, in the dialectical elaboration on the "genus-process," to the castration of the plant incapable of accommodating sexual difference. In the first instance, the *concept* of this difference falls prey to the knife of dialectics, which simplifies the entire sphere of sexuality to an oppositional relation between two sexes, the relation thanks to which each sex finds the individuality proper to it. A violent metaphysical reduction takes it for granted that only two alternatives exhaust the entire array of sexualities present, most conspicuously, in the pre-individuated state, which Freud defined as the "polymorphous perversity" of the infant and which Heidegger designated in terms of the "neutrality" of *Dasein*. The latter existential designation bears directly upon the being of plants, presumed to be similarly neutral due to its non-oppositionality, yet harboring a multitude of differences behind the façade of neutrality. The interpretation of this misleading characterization of the sexuality of *Dasein* will hold the key to the depiction of the sexuality of plants.

Apropos of the multifaceted sexual differences, presented in the guise of neutrality in *Dasein*, Derrida writes in the *Geschlecht* series: "If *Dasein* as such belongs to neither of the two sexes, this doesn't mean that its being is deprived of sex. On the contrary, here one must think of a pre-differentiated, rather a pre-dual, sexuality—which doesn't necessarily mean unitary, homogeneous, or undifferentiated [. . . but] more originary than a dyad."[69] The embodied, factical state of *Dasein* leads Derrida to this daring conclusion, mapping Freudian "pre-dual" sexuality onto *Dasein*'s purported neutrality. "More originary than a dyad," the sexuality of *Dasein* comprises sexual differences that are too subtle and too minute to be spotted through the lens of metaphysics. The same holds, *mutatis mutandis*, for the sexual being of plants. Multiple vegetal sexualities will correspond to the dispersed multiplicity at the heart of the ontology of plants that do not adopt an oppositional stance toward their surroundings. The pre-dual sexual-ontological constitutions of animals and humans are in fact indebted to this vegetal dissemination of the sexual principle.[70] While the "neutrality" of *Dasein* saturates it with sexuality to the brink (and, in any case, well beyond the confines of the dyadic relation), the indifference of vegetal sex life surpasses the logic of oppositionality and produces differences without regard to the exigencies of sameness. The front line in the fight for the liberation of sexuality from metaphysical and onto-theological constraints cuts through the being of plants.

We should remember as well that "neutrality" betokens a great deal of indifference, a certain non-involvement or non-interference of the neutral entity in events that may determine its fate. Thus in a clear allusion to the heteronomy of plants, Hegel views the seed as "an indifferent thing": "In the grain of seed [*Samenkorn*] the plant appears as a simple, immediate unity of the self and the genus. Thus the seed, on account of the immediacy of its individuality, is an indifferent thing; it falls into the earth, which is for it the universal power."[71] As we shall see in the subsequent discussions of "vegetal existentiality," the seed, entrusted to the randomness of chance and the externality of its medium (the earth), contains the possibility, indicative of its freedom, of being wasted, spread, spent for nothing. This possibility, coextensive with the seed's utter indifference, extricates it from the demands of productive or reproductive sexuality; non-teleological play, which more often than not results in nothing, is an integral part of the operativity (or, better yet, the inoperativity) of the seed.

Given that the plant's self, bound to the universality of the elements and of light, is always external to itself, its unity is at once a dis-unity, a double indifference of the light and the earth to the seeds they nourish and of the seeds to their *self*-preservation, their own fate, seeing that they have no intimate self to preserve. With this observation, we are stepping over the threshold of Derridian dissemination, where the breakdown of the unity and identity of the seed yields the multiplicity it shelters even in the singular form: "Numerical multiplicity does not sneak up like a death threat upon a germ cell previously one with itself. On the contrary, it serves as a pathbreaker for 'the' seed, which therefore produces (itself) and advances only in the plural. It is a singular plural, which no single origin will have preceded."[72] In its singularity, the seed is already a legion: whether spilled or spread, it is both one and many. Proper to the animal and vegetal modes of reproduction alike, it is at the same time appropriate to *each* animal and to *each* plant. The seed's singular plurality, on which Jean-Luc Nancy elaborates in his own thinking of community,[73] thus further specifies the sense of "vegetal democracy," where justice is understood as the aporetic combination of indifferent universality ("seed" defying the boundaries between species and even kingdoms) and attention to singularity (its appropriateness to *each* plant or animal).

The figure of the plant (which, like a weed, incarnates everything the metaphysical tradition has discarded as improper, superficial, inessential, and purely exterior) furnishes the prototype for post-metaphysical being. Plants are the weeds of metaphysics: devalued, unwanted in its carefully cultivated garden, yet growing in-between the classical categories of the thing, the animal, and the human (for the place of the weed, much like that of existence itself, is precisely in-between)[74] and quietly gaining the upper hand over that which is cherished, tamed, and "useful." Weeds will outlive metaphysics—of this we may be absolutely certain. But perhaps the greatest vegetal impurity, from the metaphysical standpoint, is the plants' overreaching to the existential domain (usually reserved for human beings alone) and their partaking of freedom, the temporal order, and wisdom (or intelligence). If, as a rejoinder to Heidegger, plants not only *are* but also *exist*, then their ethical and political status, too, will need to be revised in order to reflect their purchase on life, which has been up until now objectified under the lens of a crude metaphysical scrutiny. Vegetal existentiality, referring to the time, freedom, and wisdom of plants, will come to define the positive dimensions of their ontology.

PART II

Vegetal Existentiality

3. The Time of Plants

> Deep in the bosom of the swelling fruit
> A germ begins to burgeon here and there,
> As nature welds her ring of ageless power,
> Joining another cycle to the last,
> Flinging the chain unto the end of time.
> —Goethe, "The Metamorphosis of Plants" (A Poem)

> Le végétal tient fidèlement les souvenirs des rêveries heureuses. A chaque printemps il les fait renaître.
> —Gaston Bachelard, *L'air et les songes*[1]

The assertion that contemporary philosophy ought to take plants seriously and probe their ontological particularity does not call for a tightening of the conceptual-appropriative grasp, which vegetal being has thus far successfully evaded. The demand it poses before post-metaphysical thought is, as I mentioned in the introduction, to expose itself to the possibility, the chance, and the risk of undergoing a drastic transformation—to the point of turning unrecognizable—as a consequence of the encounter with vegetal life. A philosophy touched by the existence of plants will become livelier and more robust in the wake of such contact, but more crucially still, it will participate in the ontology of vegetation, or in what I term *ontophytology*, without projecting its own rationality upon the idealized plant.

If, on the verge of their encounter with this inexhaustibly rich "region" of being, philosophers raise the oldest question in their arsenal (the question, *ti esti*, "*What is* a plant?"), they will not yet approximate non-objectifiable ontophytology. Aristotle's answer to this query does not pursue the questioning impulse far enough; having restricted vegetal being to the double *dunamis* of plant-soul—to receive nourishment and to procreate, animating and actualizing the vegetal body—he does not take an additional step in the direction of temporalizing these capacities. In other words, the Greek philosopher does not seem to notice that the temporal character of plant-soul, one that is inherent to its capacities, involves on the one hand the continuous time of the intake of nutrients, and on the other the discontinuous time of renewal and becoming-other in whatever germinates from the mother plant. To inch closer than Aristotle to an encounter with vegetal being, we will need to rethink temporality as the mainspring of the plants' ontology, wrested from a limited and objectifying conceptual framework.

Curiously enough, various discussions of the philosophical problem of time have elevated the plant and its "life cycle" to the status of exemplarity. Germination and growth, flourishing, dehiscence, blossoming, coming to fruition, and finally fermentation and decay are indicative of the passage of time, as well as of temporalization, from Aristotle to Hegel. Nevertheless, in each case the time of the plants themselves shifts to the blind spot of these philosophical conjectures, insofar as they uncritically presume vegetal temporality to be exhausted in the process of its actualization, bringing hidden seminal potentialities to the light of full presence. What is missing from the traditional theoretical approaches is a hermeneutics of vegetal being akin to the hermeneutics of facticity Heidegger recommended as a way of apprehending the meaning of *Dasein* and by implication the meaning of being as such.

In a preliminary interpretation of ontophytology, I will advance an argument, formally reminiscent of Heidegger's conclusions apropos of *Dasein*, that the meaning of vegetal being is time. I will further isolate three key interpretations of the time of plants—the vegetal hetero-temporality of seasonal changes; the infinite temporality of growth, not precluding multiple interruptions that play an active part in the process of temporalization; and the cyclical temporality of iteration, repetition, and reproduction—each of them pervaded by the temporalizing power of possibilities that are contingent, hold themselves in reserve, resist the logic of actualization, and are linked to the impossible, the non-occurrence, the non-germination of a seed. Jointly these variations on the theme of vegetal temporality will continue to flesh out our hermeneutical approach to the ontology of plants.

Plant-Time (I): Vegetal Hetero-Temporality

Implicit in Aristotle's account of time in *Physics* is the idea that the capacities of the soul, representing the first actuality of the body they animate and indexed to types of movement such as growth, change of state, or change of position, refer, at bottom, to time, conceived as the measure of motion, *metron kineseōs* (221b8). Now, since plant-soul excels in at least two kinds of movement and change, namely growth and decay, it inherently accommodates time as its own measure, even in the restricted Aristotelian sense. Without completely corresponding to the actual content of vegetal soul, time may be understood in this instance as the quantification of the plant's ontological makeup, determined by the capacities of its particular form of life (soul). Growth and decay do not happen *in* time; they are and are not time itself, or, as Aristotle formulates it, "time is neither identical with movement nor capable of being separated from it" (*Physics* 219a1–3). The setting to work of the soul and the genesis of time are thus inextricably bound together.

At the point of encroaching upon the principle of non-contradiction, the assertion that time is and is not physical movement reveals Aristotle's inability to conceptualize temporality, keeping it strictly separate from the order of spatiality. The Heideggerian critique of a "vulgar," inauthentic, and excessively spatial thinking of temporality will accuse Aristotle of being the prime example of a metaphysical tendency to dilute the phenomena of time in the order of space. But it is by no means certain that such inauthenticity and vulgarity are detrimental to vegetal temporality, for just as plants embody sense in its finitude and materiality, so they spatially express time, illustrating the deconstructive temporalization of space and spatialization of time, or, in a word, *différance*.

The worrisome tendency in Aristotelian thought, then, is not so much the "impure" conceptualization of time hopelessly mired in the thinking of space, but rather the positing of the extra-temporal substance that endures unaffected by any empirical occurrences. Given the primacy of a stable, metaphysically safe-and-sound substance, underlying all changes and metamorphoses, time, as much as potentiality itself, is only a detour from a preexisting actuality to the actuality to come, from a past present to a future present, from a seed to a fully-grown plant. The gathering of temporality into a continuous series of "now-points"—that is to say: into a line comprised of these points—reinstates the sovereignty of the present over what is already not and what is not yet. And the image of an acorn, produced by a mature oak tree only to flourish and to become yet another fully developed tree of the same species, seals the fate of potentiality suspended between two "actualities" and therefore subsumed under the double yoke of the pure present.

While Hegel largely paraphrases Aristotle's views on time, as Heidegger has aptly observed in a famous note in *Being and Time*,[2] the German dialectician further accentuates vegetal imagery in his account of temporality. In *Logic*, the germ of a plant is said to contain its particulars, including roots, branches, and leaves, in the form of abstract potentialities that are not realized until the seed's full self-concretizing unclosing. Hegel calls this unclosing *the judgment* of the plant, alluding to the etymology of the German word for "judgment," *Urteil*, which literally means "originary division" at the outset of the processes of particularization and concretization.[3] The temporal manifestation of the plant's truth, the unfolding of plant-thinking (or "judgment") in the extended and material self-division and inner determination of this living being (as an other to itself) necessitates a transition from the potential to the actual, from the abstract to the concrete, from the merely implicit to the fully elaborated. Although these shifts are "essential" from the perspective of a "qualitative articulation" and shaping of a concrete plant, they are redundant from the metaphysical standpoint, which dissolves becoming in the actualization of pre-given potentialities: "For what is to become already is; or the becoming is this superficial movement."[4] The future, "what is to become," is already in some sense present; a mature plant, which has not yet developed by means of the qualitative articulations of growth, *is* the seed as its own not yet actualized potentiality. The superficiality of becoming (associated for its part with the superfluity of vegetal sexual reproduction) renders time itself superfluous and makes the growth of the actual plant secondary to the plant's ideal predelineation in the seed.

Readers familiar with the philosophy of Hegel will realize that the dialectical challenge, foreign to the very Aristotelianism that has inspired it, is to hold together and reconcile the metaphysical negation of becoming and the historical affirmation of growth and development. Hegel puts the plant at the forefront of this effort, delegating to it a mission on behalf of Spirit itself, so that the truth of Spirit-*qua*-plant would be relative to a particular stage of development it attains at any given moment, and would at the same time reflect the teleological actualization of the seed's hidden potentialities as a whole. The preface to *Phenomenology of Spirit* famously allegorizes the dialectical procession of *Geist* in the metamorphoses of a plant: "The bud disappears in the bursting-forth of the blossom, and one might say that the former is refuted by the latter; similarly, when the fruit appears, the blossom is shown up in its turn as a false manifestation of the plant, and the fruit now emerges as its truth instead."[5] Despite the ascription of truth to landmark events in the process of vegetal actualization, the "judgment" of the plant proceeds through a series of dialectical negations that are the organic refutations of its less developed states, refutations orchestrated, in the last instance, by the metaphysical negation of becoming itself.

Within the logic of this lopsided dialectic, time is the proper designation for the teleological transition, in the course of which the plant attains its end and final truth in fruition. While vegetal life is here confined to a set of fixed stages wherein potentialities find their actualization, time is determined on the basis of the movement from potentiality to actuality said exhaustively to explain the temporality of plants. Seeds strewn in vain and forever entombed in the earth, dried up on its surface, or indefinitely delayed in their germination; buds that wither away prior to flowering; unproductive blossoms that are not internally negated in the emergence of a fruit—all these ostensibly negative (or even purely negative) phenomena consonant with dissemination will be dismissed as mere accidents and temporary deviations from the well-charted timeline of plant growth. Similarly, Hegel would interpret anachronistic "barbarisms," of the likes of piracy, plaguing world history at the stage of its ripeness and indeed over-ripeness, as so many fleeting setbacks on the path of Spirit. From the standpoint of disseminated possibilities, however, these moments of non-fruition and non-accomplishment constitute the very temporality of time. Plant-thinking, traversing the axis of possibility–impossibility, as opposed to potentiality–actuality, will bring to the fore the temporal "truth" of vegetal life uncoupled from the teleological actualization of the seed's hidden potentialities. In so doing, it will cast off the metaphysical negation of becoming, endorse the immanently historical—which is to say, contingent—self-presentation of truth, and glimpse the elusive time of plants.

The emphasis on the role of possibilities within the temporal structure of ontophytology is in keeping with Heidegger's hermeneutic of *Dasein*, where "higher than actuality stands possibility."[6] Undertaking yet another inversion of metaphysical value hierarchies where essence is prior to existence, *Being and Time* describes, in a decidedly anti-Aristotelian way, a finite temporal existence, not supported by the enduring substance. Oriented by a multitude of possibilities, chief among them the impending futurity of death, human life does not fit into the framework of the actualization of latent potentialities. The life of *Dasein* necessarily ends in unfulfillment because immediately before the moment of death, a vast number of possibilities are not completely exhausted but continue to arise on the temporal horizon of the dying being. For Heidegger, *Dasein* meets its end in a manner that is wholly different from the ripening of a fruit; in fact, Heidegger suggests, the notion of "ending" or "coming to an end" cannot mean one and the same thing when it applies to a plant and a human being. "Ripening is the specific being of the fruit," he announces. "It is also a kind of being of the 'not-yet' (of unripeness); and, as such a kind of being it is formally analogous to *Dasein*, in that the latter, like the former, *is* in every case already its not-yet. . . . But even then, this does not signify that ripeness as an 'end' and death as an 'end' coincide with regard to their ontological structure as ends. With ripeness, the fruit *fulfills* itself [*Mit der Reife* vollendet *sich die Frucht*]. But is the death at which *Dasein* arrives, a fulfillment in this sense? . . . For the most part, *Dasein* ends in unfulfillment."[7]

Let us pay close attention to the dense texture of this passage. As the "specific being of the fruit," ripening is temporal through and through, and this confirms our initial hunch that the meaning of vegetal being is time. The core of the "formal analogy" between *Dasein* and the fruit is that at any given moment neither being fully coincides with itself, "*is* in every case already its not-yet," is not what it is. For the plant and for *Dasein*, then, the meaning of being implicates time or, more precisely, the futural modality of time (the not-yet) that resides in every present instant, dislocating its self-presence. The temporal ontologies of *Dasein* and of the plant are dependent upon the courses of their respective growths and lives, to which every interpretation of their being must pay heed.

It is only when the sense of futurity is submitted to a closer scrutiny that the hermeneutic of *Dasein* swerves away from that of vegetal life. The plant comes to its end in such a way that it is fulfilled, its *telos* accomplished without remainder in the state of ripeness, and in this it is not at all different from all other things of nature, *phusis*, whose teleology is construed on the basis of the fruitfulness of plants, *phuta*. Human beings, conversely, reach the hour of their death in radical unfulfillment denoted by the inexhaustibility of the concrete possibilities for existing still lingering before them. In other words, existential temporality with its prioritization of possibility over actuality and its effective undoing of the Aristotelian philosophy of time is but an exception that proves the general rule and leaves unchanged the essential temporalities of non-human beings. Although the being of a fruit is not fully explicable in terms of the non-existential categories "present-at-hand" and "ready-to-hand," Heidegger has silently granted that the teleological schema of time as the insubstantial shift from the potential to the actual works well for everything, save for human existence. His discussion of the ripening fruit underwrites this schema, with its adherence to the predetermined potentiality of categorial entities, to be rigorously differentiated from the open-ended possibilities of existence. The ultimate meaning of human ontology is drawn not from our life but from our relation to our own death; the plant, on the other hand, lacks such a relation, and therefore its temporality is entirely subservient to the order of life.

What we are witnessing in these statements, generally overlooked in critical interpretations of *Being and Time*, is Heidegger's subterranean conceptualization of vegetal temporality, which negatively outlines—by epitomizing everything that does not belong under the heading of existence—the central theme of the book, i.e., the idea that the meaning of the being of *Dasein* is "ecstatic temporality." The fruit's futurity, the specific sense of its coming to an end, defines the sense of its existence as well as its difference from a human being who dies unfulfilled. But doesn't Heidegger commit a serious methodological error in assuming that the teleology of ripening is proper to the being of a fruit and, more significantly, that this being, in its temporal development, may be interpreted with a view to the plant's *ownness* or *properness*? "The fruit brings itself to ripeness," he notes, "and such a bringing of itself is a characteristic of its being as a fruit. Nothing imaginable which one might contribute to it, would eliminate the unripeness of the fruit, if this entity did not come to ripeness *of its own accord*."[8] Less careful than Hegel, Heidegger does not interrogate the meaning of this vegetal self, which "brings itself to ripeness." If the plant's self, bereft of interiority, is the other—the light, or the elements, or again the environment from which it is never fully set apart—then its time, too, is not proper to it but is derivative from the other. The fruit does not actively "come to ripeness *of its own accord*"; on the contrary, it is indebted for its maturation to the rays of the sun, the minerals and the moisture obtained from the earth, let alone the artificial ripening agents such as ethylene, used to gas vegetal being out of its unripeness. The meaning of the plant's time is the time of the other, whether this "other" is a part of the organic world or a synthetically produced chemical mix, whether it pertains to the temporality of nature or to that of culture.

Historical factors also have an impact on vegetal temporality. In light of recent advances in biotechnology, it is possible to accelerate and, whenever needed, to retard the fruit's ripening—with the help of a gas 1-Methylcyclopropene or by engineering certain genes, such as LOV1, that control flowering time—and thus to harmonize the time of plants with the timing of the agro-capitalist processes of production and distribution. Land cultivation, to be sure, has always put agrarian laborers in the position of "sculptors of time," *sculpteurs de temps*, nourishing and directing the potentialities of the plants.[9] But whereas such "sculpting" still presupposed a patient expectation of the crops—which used to disclose, in the first place, the temporal core of those one who awaited them—contemporary biotechnology seems to have cut the Gordian knot of time experienced as a delay of sowing, of germination, of ripening, and of the harvest. And in so doing it impetuously disrupted the hybrid relational ontology of the human and the vegetable.

Following a classical Marxist analysis, biotechnological manipulation can be said to evince the external imposition of the commodity form onto nature in a process that manages to break and overcome the stubborn resistance of the use-value locked in the fruit, or more exactly, the independent temporality of "nature" itself. The current exploration of vegetal hetero-temporality, on the contrary, guides us to the conclusion that rather than externally impose themselves, agro-capitalist technologies internally supplant vegetal potentialities and twist them, so that they obey the demands of the economic production process.

Thanks to the fact that the being and time of the plant are to be sought in what the plant is not (that is, in its other), its potentialities are left vacant for infinite appropriation by anything or anyone whatsoever. The commodification of the plant's time, nearly nullifying the wait for its development toward ripeness, parasitically exploits the heteronomy of vegetal temporality when commodity logic turns into the plant's other, and finally into the source of its meaning. Ontologically, what allows us to capture the potentialities of the fruit is the realization—whether conscious or not—that it never autonomously appropriates its time and self. Standing in for the other, in which the plant's being accomplishes itself without reflectively returning to itself, capital eclipses the sun and the power of the nutrients contained in the earth.

The hothouse is perhaps a less dramatic example of the human mastery over the time of plants, our success in manipulating their very being by altering the environmental conditions responsible for the process of fruition. In any event, techno-cultural and economic phenomena do not negate a preexisting "natural" condition but interject themselves into the place of the plants' other, indifferently occupied by any force exerted on vegetal being and so shaping ontophytology. Coming to fruition and ripening are in sync with the change of seasons only because the time of the other immediately spells out the meaning of vegetal temporality (even though the reverse influence of the time of plants on the human understanding of environmental time is also strong: etymologically, *season* comes from *serere*, "to sow," which is, incidentally, the root of *semen*, *dissemination*, *seminal*, *seminar* . . .).

By emphasizing the *internal* supplanting of vegetal time I do not mean to suggest that the positing of the human and capitalist temporalities in the place of the plant's hetero-temporality is nonviolent. Mastery over an entity's time is immediately translatable into mastery over its being; a telling allegory of violence, the interjection of capitalist temporality into the place of the other of vegetation ultimately transforms the other into the same, subsuming the temporal being of the fruit to the hegemonic time of capital's brand of "growth" and "maturation," in accordance with the demands of value's self-valorization. In an effort to resist this subsumption, it would be futile to appeal to the redemptive implications of the fruit's natural ripening, since the potentialities of the plant are never completely its own but are contingent upon the fragile situation of its coming to fruition. The only effective resistance imaginable would be one that insists on the non-synchronicity, the asymmetry, and the non-contemporaneity of human and vegetal temporalities and that releases the time of plants back to the contingency of the other, spelling out its meaning, time and again, according to the singular context of its embeddedness. Hence, the locus of resistance would be the time of plants not measurable in human terms, that is to say, in terms of the movements proper to human beings and their kind of soul.

In *Matter and Memory*, Bergson underlines the unbridgeable divergence between the lived duration of human consciousness and the rhythms and vibrations defining the purely objective time of the physicist: "The duration lived by our consciousness is a duration with its own determined rhythm, a duration very different from the time of the physicist, which can store up, in a given interval, as great a number of phenomena as we please . . . . In reality there is no one rhythm of duration; it is possible to imagine many different rhythms which, slower or faster, measure the degree of tension or relaxation of different kinds of consciousness and thereby fix their respective places in the scale of being."[10] The pulsations of vegetal temporality are often imperceptible to a conscious human observer, because even when they share a physical space, the two beings do not live in the same homogenous time but are non-contemporaneous with one another. Beneath the Husserlian active synthesis of memory and passive synthesis of perception, uniting disjointed instants into a temporal sequence, lies the still more passive organic synthesis of metabolism, constituting a significant portion of the plant's time. Since we cannot accompany, continuously, the temporality of vegetal growth (whether in plants or in ourselves!) that passes unnoticed below the threshold of human perception, it will appear to us, in any given period of lingering with the plant, that its time is virtually nonexistent or, in Aristotle's terms, that its movement is immeasurable. Paradoxically, there needs to be a break or a rupture in our temporal approach to the plant—for example, coming back home after a long period of absence—in order for us to register the passage of time expressed in the spatial increase in the plant's stem, leaves, and so forth.

"Vegetal time: they [the plants] always seem frozen, motionless. You turn your back for a few days, a week; their posture is more clearly defined, their limbs have multiplied. There's no mistaking their identity, but their form is more and more realized."[11] To turn one's back to the plant is to be no longer conscious of it, disrupting the intentional comportment that relentlessly and futilely projects human expectations, rhythms, and durations onto vegetal and all other types of being. Distraction from the vegetal other dispenses its otherness back to the time of life qualitatively different from human existence. In the gap, formed as a result of interrupting the relation between human and vegetal temporalities, the time of the plant finally discloses itself, albeit not as pure temporality but as a relative spatial increase: a clearer definition of its posture and a multiplication of its stems and shoots. But as soon as the conscious attitudes of weighing, measuring, and comparing turn the spotlight back on the plant, vegetal temporality, untranslatable into the intervals of duration familiar to human consciousness, dissolves into vegetal spatiality.

Not only are the existences of plants and human beings noncontemporaneous with one another, but each also does not live in the same time with itself. On the human side, the rhythms of consciousness are exquisitely varied, oscillating between, say, the distended duration of boredom and the condensed waves of intense joyfulness. The supplement to Husserl's phenomenological description of the heterogeneity of inner time consciousness is Heidegger's existentialism, according to which human beings are never identical to themselves, since their ecstatic temporality consists of a constant projection of themselves into the future, a temporal dissociation from themselves, and a lagging behind this futural self. It is because ecstatic temporality makes them noncontemporaneous with themselves that humans are able to posit themselves over and against their environment, to determine themselves by opposing themselves to their place (Hegel), to live "out of season" (Nietzsche),[12] or discursively to establish psychic interiority by setting it apart from the rest of the world (Levinas). The name for such noncontemporaneity is *culture*, but that is not to say that what culture opposes, rebels against, and objectifies, even as it leads the afterlife of its repressed or disavowed object, persists in the form of a pure identity.[13]

The plant too is not contemporaneous with itself, in that it is a loose alliance of multiple temporalities of growth—some of its parts sprouting faster, others slower, still others decaying and rotting—and in that it does not relate to itself, does not establish a self-identity. Its non-synchronicity with itself is not the product of an excess overflowing a unified self, as in the case of the human, but is instead an outcome of the absence of identity that forces it to obey the law and the time of the undifferentiated other, assigning to vegetal being the qualities of heteronomy and hetero-temporality.

At this point we find ourselves at an ethical junction, best envisioned with reference to the philosophy of Emmanuel Levinas. *On the one hand*, the emplaced existence of vegetal being is a distinguishing mark of the imperialism of the same, which is palpable in the acts of putting down roots and claiming a place for oneself and which readily lends organic metaphors to a socially and politically conservative orientation. It is not difficult to recognize allusions to Heidegger's imprudent romanticization of bucolic life in Levinasian criticism of autochthony, of "the virtues of being warrior-like and putting down roots, of being a man-plant, a humanity-forest whose gnarled roots and trunk are magnified by the rugged life of a countryman."[14] This extreme attachment to the place, the resources of which are not offered to the other, precludes hospitality and, to an equal extent, facilitates individualism: "What is an individual, a solitary individual, if not a tree that grows without regard for everything it suppresses and breaks, grabbing all the nourishment, air, and sun, a being that is fully justified in its nature and its being? What is an individual if not a usurper?"[15] Vegetal being is here treated as the epitome of unethical existence, one that is self-referential and appropriative, as well as blissfully indifferent to the needs of the other.

Were Levinas to ask what appropriation means for a being devoid of interiority; whether the exclusion of the other is possible in the condition of absolute exposure and vulnerability; and to what extent the plant is solitary in itself and in the place where it grows—were he to reflect on these issues, would the correlation he draws between the tree and the individual have been secure? Even more questionable is his other assumption, namely, that the plant's embeddedness in a place (metaphorically, in the concreteness and parochialism of paganism) necessarily implies that it is insulated from the ethical order of time. Does temporality require a stepping out of place, a spatial negation of space, for Levinas as much as for Hegel, in a certain ritual of purification, transcendentally inaugurating a philosophy of temporality that is "authentic" because no longer affected by spatial thinking? If it does, then the time of plants is an absurd proposition indeed.

Further down on the same fertile page of "Place and Utopia" Levinas admits that one may assume ethical responsibility only in a particular place, with which this responsibility should not merge and to which it should not be limited. The utopian drive, tending toward displacement and uprooting, is a mere excuse for absolving oneself of responsibility: "One can uproot oneself from this responsibility, deny the place where it is incumbent on me to do something, to look for an anchorite's salvation. One can choose utopia." The non-place is at least as detrimental to an ethical orientation as the total enchainment to a place wherein one is rooted. Like paganism, worshiping the gods of concrete locales, the abstract universalism of science, indicative of utter displacement, is incapable of responding to the call of the other. What is required instead is the structure of transcendence within immanence (which Levinas wants to tease out of Jewish monotheism) and therefore the necessarily impure temporalization of space.

Still, the accusation (reflecting a charge leveled against vegetal being and stemming from self-righteous monotheism) of paganism as unethical is overly precipitous. Distinct from transcendence and placelessness that *post factum* exaggerate one half of the ethical orientation, immanence and the place are the indispensable conditions of possibility for ethics. Vegetal being and the capacity for nourishment it has honed are the centerpiece of the material need of the other who appeals to the I, converting it to ethical subjectivity in the here-and-now of its existence. Nor should the place be conflated with space, for it is in the former that the latter is temporalized. The plant's inability to negate its place does not leave it bereft of time but weds it to hetero-temporality in the fullness of its immersion in and anchoring to a place.

On the other hand, in the works of Levinas, temporalization always involves a relation to the other, the very relation that (we might add) is at the core of vegetal ontology. "Relationship with the future," Levinas writes in *Time and the Other*, "the presence of the future in the present, seems all the same accomplished in the face-to-face with the Other.

The situation of the face-to-face would be the very accomplishment of time."[16] Although the plant is not separate enough from its environment to find itself in a face-to-face situation with anything or anyone whatsoever, in its very being it exemplifies the tenet that time is bound to alterity. The hetero-temporality of vegetal existence is the most telling instantiation of the ethical injunction for openness to the other. The plant's future is entirely contingent on alterity when it comes to the process of ripening, the possibilities of flourishing and withering away, and so forth.

Implying neither a conscious choice nor the impassiveness of inanimate objects, the plant's sheer exposure in space and in time bespeaks what Levinas terms "passivity more passive than all passivity," the feature of an ethical comportment "in its antecedence to . . . freedom, its antecedence to the present and to representation."[17] Before activity, before a conscious orientation, and before attachment to the present, the time of the other determines the being of the ethical subject as much as that of the plant. Responsibility, in the normative and calculative senses of the word, pales in comparison to its semantic association with responsiveness and exposure to the other as "the very accomplishment of time" in vegetal being.

Plant-Time (II): The "Bad Infinity" of Growth

Vegetal time passes in qualitatively distinct modes and rhythms. In addition to hetero-temporality, the potentially infinite movements of growth and efflorescence further specify the time of plants (or more precisely, the meaning of this first temporal modality) as the way they tend to their other without measure, without limit, without term, and without ever reaching their final destination: "Thus, a mighty tree reaches the sky, dwells there, and extends itself without end [*Ainsi, l'arbre puissant atteint le ciel, s'y installe, s'y prolonge sans fin*]."[18] Such monstrous growth and immoderate proliferation, whose possibilities are, *stricto sensu*, never realized, have always been unspeakably terrifying for philosophers, who in one way or another have busied themselves with, on the one hand, establishing the "proper limits" for desire, reason, life, or action, and, on the other, with setting up conceptual police authorities to safeguard these limits against potential transgressors. The plant's "endless growth outwards,"[19] its total externalization, the "infinite distances of the floral world [*Unendliche Ferne der Blumenwelt*],"[20] and infinite temporality are anathema to the basic orientation of philosophy toward completion and perfection. Whenever a metaphysical philosopher speaks of plants at all, it is with the purpose of taming their proliferation and of appropriating their time, measuring it, and declaring it deficient in keeping with this measure alien to human beings.

Take vegetal desire, for instance. Whether or not the plant is a desiring being, it experiences no satisfaction when it exercises the only capacity of its soul for nourishment. In an already cited fragment of *The Will to Power*, Nietzsche calls the drive for nourishment, which is the epiphenomenon of the will to power, "insatiable,"[21] while Novalis attributes to the life of plants an "uninterrupted eating and fecundation [*ein unaufhörliches Essen und Befruchten*]."[22] Here both echo Hegel, who writes on the subject of the plant's nutrition that "it is not an interrupted process but a continuous flow" and that "air and water are perpetually acting on the plant; it does not take sips of water."[23] Since the plant concentrates its entire being in the act of nourishment, it has no time to engage in other activities and therefore, on this view, it has no time at all. Time without a finite term, without spacings, ruptures, or terminations; time passing as a "continuous flow"; the time of nourishment without respite dissolves, like the very eternity it mimes, into pure space, even though it arises from the activity—*the only* activity—of the vegetal soul. The animal, in contrast to the plant, will gain time by virtue of the intricacy of its digestive system, which allows it gradually to expand energy and to be released from the constant attachment to the source of nourishment. No wonder then that one of the physiological features of the human being Hegel admires the most in *Philosophy of Nature* is the length of its digestive tract—directly proportional to the time of life freed from the exigencies of feeding! Beyond the kingdom of plants, dialectical time first arises in and as a suspension of the immediate nutritive activity, mediating and sublimating nourishment within the organism, thereby liberated for other pursuits.

But what exactly do we mean when we say that a living being spends all its time on one activity? Does the heedless immersion of the plant in nourishment and growth not betray its utmost non-conscious attention to life itself? And isn't it an unjustified argumentative leap to assume that when animals pursue activities other than nourishment, they leave behind, if only for a moment, this basic stratum of life? If an animal or a human being is able to refine, delay, and internally mediate the process of nutrition, this does not imply that it has successfully evaded this process in anything it does or, in the case of a human being, *thinks*. Time itself is the sublation of nourishment, which, crude as it may be in plants, is never instantaneous to the extent that it is a mediation of the nutrients and the nourished body. Vegetal temporality persists in us, despite being modified, internalized, and consequently concealed, creating an illusion of self-sufficiency.

Like the seasonal existence that largely defines the hetero-temporality of vegetal life, infinite growth, immoderately aspiring toward the other, has neither a beginning nor an end.[24] Such growth resonates with Levinasian metaphysical desire, divorced from "the deceptions of satisfaction" and "the exasperation of non-satisfaction," one that, in its inordinateness, "desires beyond everything that can simply complete it."[25] Vegetal insatiability in the face of a continuous flow of nourishment and human metaphysical aspiration, designating an ethical approach to alterity, invalidate a straightforward association of desire with lack; in fact, the positivity of the two desires harbors the meaning of time itself. On the hither side of actualization, the plant and the ethical subject dovetail in becoming what they are only *on the way to* and *for the sake of* the other. Their infinite, interminable, permanently incomplete passage to the other, where "Not enough!" signals that the activities associated with being-for-the-other must go on, *is* ethical and vegetal time. Yet even here the uninterrupted continuity—or the pure presence—of the flows of nourishment and desire is nothing more than a theoretical fiction. Ethics and growth entail various possibilities of self-interruption, making possibility itself possible and precluding both a return of the subject to itself in the condition of enjoyment and a merely quantitative augmentation of the growing body that would not lead to its qualitative transformation.

The first self-interruption of the time of vegetal growth is discernible in the periodic pace at which it proceeds. Goethe's theory of the metamorphosis of plants accentuates the *rhythm* of expansion and contraction, so that "the organ that expanded on the stem as a leaf, assuming a variety of forms, is the same organ that now contracts in the calyx, expands again in the petal, contracts in the reproductive apparatus, only to expand finally as the fruit."[26] But a rhythmic movement necessarily incorporates into itself ruptures and discontinuities, such that "'rhythm' has its proper moment only in the gap of the beat that makes it into rhythm."[27] There is always a spacing or a delay, however miniscule, between the moment of expansion and that of contraction, a spacing that temporalizes the continuous flow of nutrients and determines its direction in a quasi-musical fashion, emphasizing nature's own cadences, beats, measures, and melodies: the pace or rhythm of *phusis* as emergence.

It is thanks to this "gap of the beat," thanks—also—to the internal interruption of the spatial flow, that, in keeping with Derridian *différance*, the time of vegetal growth is not reduced to space. In its turn, deconstruction frequently casts *différance* in terms of dehiscence, the maturational opening of a plant structure, be it a bud or a fruit: "As in the realm of botany, from which it draws its metaphorical value, the word [dehiscence] marks emphatically that the divided opening, in the growth of a plant, is also what, *in the positive sense*, makes production, reproduction, development possible."[28] There is no growth, no matter how "uniform" and "simple,"[29] outside a divided opening, a *différantial* interruption of what has been moving by inertia, and there is no time without a fissure in the relentless contiguity of the spatial order: the seemingly mechanical rhythm of vegetal growth contains discontinuity right in the midst of continuity and time in the midst of spatial augmentation.

Goethe and Hegel have noted another temporalizing self-interruption in the metamorphosis of plants. The activation of the reproductive function—which, as the reader will recall, is the second major capacity of the vegetal soul in Aristotle—prompts a slowing down of the plant's quantitative growth. The vegetal soul is split against itself, in that it is able to valorize one of its capacities only at the expense of the other: the more the plant grows, the slower it reproduces, and vice versa. And time is nothing but the positive "effect" of such splitting. Flowering, "this nodulation which arrests growth,"[30] responds to the same animating impulse as growth itself, which strives toward its inorganic other (the sun). Why? Because reproduction still tends toward alterity, this time by *engendering* the other. The being-toward-the-other of growth and the becoming-other of reproduction fissure the time of plants, which proceeds in and as a series of self-interruptions.

At the confluence of the ethical and the botanical, fecundity, this quintessentially vegetal capacity, stands for a ruptured continuity and a personal transcendence in Levinas's philosophy, as much as it underlies the "infinity of time," the continuous approach to the other both with and without me: "In fecundity the tedium of this repetition [of the reiteration of the I] ceases; the I is other and young, yet the ipseity that ascribed to it its meaning and its orientation in being is not lost in this renouncement of the self. Fecundity continues history without producing old age."[31] The time of an individual being remains finite (fecundity does not dispense with the inevitability of death) even as it fulfills the promise of infinity, of the continuation of existence in the other, who stands in the place of the individual progenitor. The art of self-interruption is the secret nexus of ethics, orienting me toward the other by urging me to suspend my own enjoyment for the sake of metaphysical desire, and of life, orchestrating the engenderment of the other at the expense of limitless growth and self-augmentation.

The internal interruptions of vegetal temporality are the cipher of finitude at the heart of infinity, the traces of ageing on the ostensibly immortal body of the plant. The sobering acknowledgment of this body's precariousness, fragility, and temporality goes a long way toward preventing the unethical treatment and abuses of vegetal life, which is usually taken to be an eternal reserve that, try as we may, cannot be depleted. But it also makes the love of plants possible. As Bergson states, "It is easy to argue that the tree never grows old, since the tips of its branches are always equally young, always equally capable of engendering new trees by budding. But in such an organism—which is, after all, a society rather than an individual—*something* ages, if only the leaves and the interior of the trunk. . . . *Wherever anything lives, there is, open somewhere, a register in which time is being inscribed* [Partout où quelque chose vit, il y a, ouvert quelque part, un register où le temps s'inscrit]."[32] Only temporal, finite, mortal beings may be the recipients of love; to attribute these qualities to a plant is to confirm that it is potentially *lovable*. Time in all its finitude is stamped on the collective and dispersed body of the plant. Vegetal *différance* inscribes the time of plants right onto the spatial register of material sense, just as it gathers, without synthesizing, the most indefinite ("something," "anything," "somewhere") and the most defined (the tree), eternal youth and irreversible aging. Or perhaps time is not inscribed but, in a word Jean-Luc Nancy has coined, *ex-scribed* on the vegetal body, which, on the hither side of the metaphysical distinction between interiority and exteriority, marks time in a peculiarly geometrical style, by the accretion of "rings," those symbols of eternity and indicators of the tree's inexorable aging. Time does not in fact preexist such ex-scription but derives from the *différantial* "opening" of the register, wherein it leaves its traces over and over again.

Faithfully following and to some extent embodying the annual cycles with their incessant alternations of decay and regeneration, the tree permits us to appreciate repetition not as the "tedium" to which Levinas has confined it, but as the mute affirmation and reaffirmation of existence in all its finite materiality.

Plant-Time (III): The Iterability of Expression

The time of growth is both linear, insofar as it tends to infinity, and cyclical, if its sense extends to the root meaning of *phusis* as "growth not only of plants and animals, their arising and passing away taken merely as an isolated process, but growth as this occurring in the midst of, and permeated by, the changing of the seasons, in the midst of the alternation of day and night, in the midst of the wandering of the stars. . . . Growing is all this taken together as one."[33] In an exaggeratedly metaphysical theoretical gesture of grouping together all living beings, Heidegger obfuscates the particular temporal mode of being of plants. What he overlooks in a gloss on "plants and animals," situated in this text on the same ontological plane, is that the "arising and passing away" of vegetal beings is never "an isolated process" but rather is in tune with and subject to the gathering of growth under the semantic aegis of *phusis*.

Redoubling the global movement of growth "taken together as one," plants precipitate the non-dialectical coincidence of the singular and the universal at the point where the cyclical time of nature (the changing of the seasons, the alternation of day and night) intersects with the cycles of vegetal growth (the budding and shedding of foliage, the opening and closing of a flower). The plant metonymizes this most encompassing circle and reflects it as though in a miniature mirror, where the circulation of sap, representing the plant's "seminal power" (*la puissance séminale*), obeys the circularity of seasonal changes or indeed the rotations of the Earth.[34] Far from a merely mechanical reproduction that would testify to the provenance of plants from the inorganic world, this repetition at once avows every single moment of vegetal existence and confirms its unquestioned allegiance to the other.

This leads us to the hypothesis that the plant, with its non-conscious affirmation of repetition, prefigures the affirmative movement of the Nietzschean eternal return, with its acceptance of the perpetual recommencement of life. When, in the spirit of Nietzsche, Michel Onfray refers to floral time as "cyclical," he contrasts this vegetal acquiescence to the heroic stance of humanity, incapable of learning from the plants' temporal existence.[35] Repetition befalls purely active agents as a tragedy (and, according to Marx, as a farce, provided that it is a repetition of repetition); it is a mythic force, to which the subjects, the fictitious masters of their own fate, are subjected without ever giving their assent. But the real tragedy happens in the guise of the futile struggle against repetition, against life itself, and against the very construction of the subject around certain reiterated habits, practices, and discourses. Despite the traditional association between mechanical reproduction—the symbol of the merely external compulsion—and death, repetition achieves the exact opposite of deadening; it stakes out the being of everything that is living, that is to say, of everything that exists thanks to its inability to maintain a static identity to itself. There is no life without iterability: the possibility of recommencement, built into the dis-organized vegetal body, as well as into the bodies and "souls" of human beings.

One of the most emblematic iterable parts of the plant is the leaf, upon which we have already focused in our discussion of an essentially superficial ontology of plant life and to which we will keep returning. The leaf: an ephemeral register for the inscription of vegetal time as the time of repetition, a register not archived but periodically lost and renewed, such that these losses and renewals themselves make up the temporal, temporalizing trace imprinted on it. As Deleuze puts it in *Difference and Repetition*, "repetition is a necessary and justified conduct only in relation to that which cannot be replaced,"[36] even something as trivial and insignificant as a needle of the pine tree I am observing from my window. Only the irreplaceable, only absolute difference *in itself* may be repeated—this aporetic maxim is alien to the discourses of sustainability (in truth, to the entire logic of "sustainable development"), incapable of recognizing anything in the so-called natural environment as irreplaceable, and, as a consequence, positing the idea of equivalence between the "resources" used up in the past and those meant to be their substitutes in the future. The temporal-ontological marker of vegetal life is, on this reading, transformed into a justification for the abuse of this life. Conversely, Deleuze argues that "repetition is the thought of the future,"[37] which, in maintaining fidelity to that which cannot be replaced, reiterates and projects past difference. Neither an autistic singularity, worshiped by radical empiricists, nor a vacuous generality, revered by adherents of idealism, difference *in* repetition, as well as the repetition *of* difference as difference, is the mold for the time of vegetal growth encrusted in the larger circle of *phusis* it reflects.

We are now better equipped to interpret Francis Ponge's admiration for the propensity of plants to "repeat the same expression, the same leaf, a million times" and, "bursting out of themselves," to produce "thousands of copies of the same . . . leaf."[38] Rather than attribute an idealizing function to repetition in vegetal existence, Ponge glimpses the language of plants in its temporal and concrete unfolding. Each leaf is a repeated expression, a reiteration of the difference that "is" nature, a reaffirmation of the finite, material sense of existence it epitomizes. If it takes time for this semantic-living content to be expressed, over and over again, this is not because a certain encrypted message is still on its way to an unknown addressee. Devoid of a fixed destination, countless copies of expressions-leaves are disseminated as the concrete self-representations of vegetal life. (Deleuze calls such representation "orgiastic," noting that it "makes things themselves so many expressions or so many propositions.")[39] As living representations, the leaves may be grouped with the other fragments of the Leibnizian *mens momentanea*, the body cast in terms of a "momentary mind" perceptible, Deleuze holds, in repetitions.[40] In plants, bursting out of themselves with every new copy of the leaf, nature stands out of itself—or else, ecstatically announces itself and temporalizes itself.[41] Acts of repetition do not clarify anything whatsoever, do not consolidate or crystallize the structure of meaning they carry, but simply affirm, with renewed energy, the sense of vegetal existence, a sense which fuses with this very existence in all its heterogeneity and finitude. Besides furnishing the non-transcendental conditions of possibility for the time and the being of plants, these acts account for the genesis of the plants' temporal-ontological meaning or, in other words, for the hermeneutics of vegetal life.

If Deleuze is right in claiming that only the irreplaceable can be repeated, the iterations of the leaf we have been monitoring up until now do not contradict Leibniz's fascination with the distinctiveness of every single leaf, illustrating his "principle of the identity of indiscernibles," the idea that, if two things (e.g., two leaves) were exactly identical, they would have been one and the same thing.[42] As Hegel observed with a great measure of irony, Leibnizian difference has nothing to do with the empirical or ontic dissimilarity of two discrete things—a misunderstanding that led "the cavaliers and ladies of the court, as they walked round the garden" to search in vain for "two leaves indistinguishable from each other, in order to confute the law stated by the philosopher"[43]—but, more abstractly, with the implicit ontological difference that dialectically sets up each thing as what it is in opposition to what it is not. Still, this solution, inaugurating the repetition of the irreplaceable as a temporal and ontological principle, runs the risk of falling into the trap of metaphysics, if, in reacting to the "vulgarity" of the empirical approach, it idealizes difference raised to the status of a transcendental principle. Plant-thinking must perform a delicate balancing act of avoiding both crass empiricism and metaphysical excesses, but to do so we ought to turn for guidance to the Derridian notion of iterability.

Derrida reminds his readers that "'iterability' does not signify simply . . . repeatability of the same, but rather alterability of this same idealized in the singularity of the event. . . . There is no idealization without (identificatory) iterability; but for the same reason, for the reasons of (altering) iterability, there is no idealization that keeps itself pure, safe from all contamination."[44] The possibility inherent in iterability—of repeating the same with inexorable alterations in each singular contextual instance—underwrites practices of signification as much as the proto-inscription of nature and in particular the ontology of vegetal life. Comparable to every word or concept that will carry slightly different semantic overtones depending on the singular event of its enunciation, the shape of every leaf will vary with each repetition of the material expression of the being of plants, the expression that "is" this very being. It is thus one of the exigencies of plant-thinking to revise the metaphysical-semantic homology of signs and seeds (*sema*), which extends from Plato's dialogues all the way to the deconstructive emphasis on the dissemination of meaning. As a result of this revision, the essential superficiality and non-originality of the sign as a mark or a trace would find its reflection not only in the seed but also in the iterations of the leaf, the presumed support for the inscription, which is, at the same time, the content of the plants' proto-writing.

In vegetal life as much as in signification, time performs the ontological labor of de-idealization, insofar as the repetition of the same is altered in "the singularity of the event." This event is a cleft, a break, a dehiscence (let it be reiterated that Derrida is extremely attentive to the botanical derivation of this last term) around which identity gathers itself, impotent to form a simple undivided unity. In this impotence, the thought of identity replicates the failure of vegetal growth to ward off that which interrupts it, often from within. Not only is the same—a leaf, a word, a semantic unit—repeated each time anew, becoming other in each repetition, but the temporal dimension of the future is also left ajar, maintaining the possibility of iteration alive. Hence the stress on iter*ability*.

For the human observer, temporalizing vegetal possibilities, well in excess of the closed Aristotelian circle of potentiality and actuality, hold the promise of future regeneration as much as the memory of past springs: "The plant faithfully keeps memories of happy daydreams. Every spring, it gives them a new birth [*Le végétal tient fidèlement les souvenirs des rêveries heureuses. A chaque printemps il les fait renaître*]."[45] The event of vegetal iteration stretches between the past archived in the temporal register of the plant—which is to say, ex-scribed in the materiality of its being—and the possibility of future regeneration. In the vegetal mode of ec-static existence, past thrownness, commencing with the literal throw of the seed, and future projection, bespeaking the plant's growth or non-conscious intentionality, spell out the finite temporality of its life. The freedom of plants, indebted to disseminated possibilities, and the sagacity of plants, predicated on their embodied, living, non-conscious intentionality, will further specify the meaning of their being conceived in terms of vegetal existential temporality.

4. The Freedom of Plants

Flowers are free beauties of nature. Hardly anyone but a botanist knows the true nature of a flower, and even he, while recognizing in the flower the reproductive organ of the plant, pays no attention to this natural end when using his taste to judge its beauty.
—Immanuel Kant, *The Critique of Judgement*

Pour nous libérer, libérons la fleur.
—Francis Ponge, *Nouveau Nouveau recueil*, 1967–1984[1]

Across a monolithic metaphysical tradition—subtending the thinking of the free will as much as the political ideal heralded by the French Revolution, the Kantian and the Hegelian equation of freedom with the capacity for self-determination, Isaiah Berlin's thematization of "negative" and "positive" liberties, and most recently, the existential-ontological "freedom unto death," to mention but a few vivid examples—silent consensus reigns: the possibility of freedom is foreclosed to the plant. What can we glean about freedom from a being presumably devoid of selfhood, whose roots tether it to the earth, making it into a veritable symbol of stupor and immobility? If the plant has no self, let alone an individuated "will," how should it be taken up into the discourses surrounding this idea, itself indefinite and ambiguous because largely free of conceptual determinations?[2]

Negatively and obliquely, vegetal beings will circumscribe the form of freedom (both the physical curvature proper to it and its metaphysical openness to being as such) with reference to their own geometrically arranged, symmetrical, "crystalline" structure and environmental fixity. Or perhaps they will gain a foothold in the vacant sphere of freedom, as it were, through the backdoor, thanks to the assertion that to be free *is* to be without a head (*ergo* without consciousness), without worries and concerns—to be indifferent, to be a plant. Or more positively, they will broach the subject of freedom through those cracks in the metaphysical edifice cropping up on the margins of the tradition, where the plant, specifically the flower, suddenly flourishes, as in Kant's *Critique of Judgement*, or refuses to blossom, as in Schiller's aesthetics. In one way or another, the ingress of the plant into the discursive terrain of freedom will transform these discourses from within, liberate them from the dead weight of metaphysics, and thus remedy the "fallaciousness" of this overused philosophical keyword.[3]

The Shape of Freedom

According to one of its significations, "freedom" refers to the possibility for an entity to be otherwise than what or who it is, which is why proponents of German Idealism and Romanticism saw in the faculty of imagination (permitting us "to be otherwise" by mimetically putting ourselves in the shoes of other beings, whether human or not) the *locus essendi* of freedom. It is therefore an ontological term, one that pertains to the being of a free being that is capable of not being as it is. For the purposes of the current chapter, the question is: Does this barest of standards apply to plants and their mode of being? Can vegetal beings be otherwise than they are?

Early on in *Physics*, Aristotle seems to preclude this very possibility, when he claims that among other things, plants "exist by nature": "Some things exist, or come into existence, by nature [*esti phusei*]; and some otherwise. Animals and their organs, plants [*phuta*], and the elementary substances—earth, fire, air, water—these and their likes we say exist by nature" (192b8–11). On the first reading, substantiated by the remark that "nature is etymologically equivalent to *genesis* and is actually used as a synonym for it" (193b14–15), the attribution of the plants' existence to "nature" means that it is impossible for them to be otherwise, since they are entirely and originarily determined from birth—"genetically determined," as we would say nowadays—and therefore ontologically unfree. It appears that vegetal beings are slotted into a narrow niche within the grand order of nature, wherein they would be prevented from determining themselves by themselves and from exceeding the province of their existence. Isn't this after all the meaning of vegetal heteronomy in the time of plants?

Although the Latin *na-tura* is similarly derived from the *na-* of *nascor* ("to be born"),[4] both "nature" and "origin" conceptually distort the Greek *phusis* and *genesis*. In the aftermath of translating philosophical vocabulary into Latin, one loses sight of the meaning of "nature," which used to point to the internal determination of the natural entity, the source of movement and of being at rest inherent to it as such.[5] For the Greeks, self-animation is precisely the hallmark of all natural beings, invested with a certain degree of freedom in comparison to works of art (*tekhnē*), which are indebted to external causes for their movement and, indeed, origination. Existence "by nature," in this semi-forgotten way of thinking, comes close to denoting existence by itself, not in the sense of its autonomous production *ex nihilo* but as an emergence out of itself of growth (one of the four Aristotelian types of movement), taken as a whole under the heading of *phusis*. And while plants are largely at the mercy of external elements, they are not dependent on their other in the same way as manufactured artifacts are.

Recall, moreover, that the plant is not just a thing in nature, but the miniature mirror of *phusis*, a synecdochic instantiation of universal growth and its refinement in the faculty of the vegetal soul, the "lowest" and at the same time the most encompassing potentiality of all living beings. The plant–nature synecdoche corroborates the sense of vegetal freedom, shedding a different light on the expression "to exist by nature," which in the case of the plant implies "to exist by itself." If for a vegetal being existence by nature signifies, at bottom, existence by itself, then it is, in and as itself, ontologically free, liberated in the core of its being. Now, Aristotle further buttresses this implicit synecdoche by raising reproduction—the capacity of the vegetal soul *par excellence*—to the level of a criterion for telling natural things apart from products of *tekhnē*: "If, then, it is this incapacity for reproduction that makes a thing art and not nature, then the form [*morphē*] of natural things will be their nature" (193b8–13). A prologue to the theory of fourfold causality, the Aristotelian morphology of natural things identifies their souls with their forms, which define their natures.

But vegetal soul, concentrating in itself the capacities for reproduction and growth, is exceptional in that it becomes the (embodied and finite) form of these forms; in fact, it stands for the very nature of "natural things," as distinguished from artifacts. Apropos of freedom, the ancient synecdoche suggests that the vegetal principle endows its bearer with the greatest facility in being otherwise than it is in its most recognizable shape of a plant, because, though it retains the vestiges of plant life, this principle stimulates the growth and reproduction of other living beings, including animals and humans, who are likewise growing beings. Adjusting to the forms of all natural things, the *morphē* of plant-soul is extremely elastic, to the point of indefiniteness—a quality that underwrites its freedom of multifaceted involvement with the other beings that exist by nature.

We have already observed how the relative paucity of purpose or end (*telos*) in the life of plants has, in the aftermath of *De plantis*, been attributed to the ontological defectiveness, or incompletion, of vegetal being. The indefiniteness of their "purpose," however, leaves vacant the space of freedom and removes plants from the circumscribed and inferior position in the teleological chain typically allotted to them. Even in Aristotle's text, where it is by no means "a matter of chance [*tukhē*] what comes up when you sow this seed or that" (199b14), the nascent teleology of vegetal life is malleable enough to blend with the ends of other beings, for in plants the *teloi* are not "elaborately articulated [*diarthrōtai*]" (199b11). The non-articulation of their ontological purposes, indicating—besides the rudimentary stage of life's differentiation at which the plants find themselves—the inarticulateness of the philosophical discourse on plants that makes this pronouncement, partially disengages them from the effects of metaphysical (and later on, conceptual) violence of classifying nature and submitting it to a unified order, principle, and beginning: in a word, *archē*.[6] Beneath a plurality of medicinal, nutritive, sheltering, and other ontic purposes projected onto plants, philosophy futilely struggles to espy the objective *telos* of vegetal life. Thus unfixed as regards its final purpose or end, the plant attains its freedom as a consequence of being banished outside the strict teleology of nature and muddying the categorial boundaries between various classes of living beings that depend for their survival on the basic features of the vegetal soul.

There is nevertheless a grave obstacle on the way out of the metaphysical framework: the discovery of the most plastic and in this sense the freest form of life (vegetal life is at the same time exclusive to plants, and stands for life as such) pivots on the metaphysical master-distinction between natural and artificial things, between the originary and the derived, the living and the dead. It is insufficient to extract plant life from the hierarchy of ends without freeing it from the stricture of pure living, in a gesture that would wreak havoc on the neat Aristotelian segregation of natural beings from products of art. Such is the vocation of Derridian deconstruction, which bears witness to the fact that the "effort to *render* the flower can only fail. . . . The flower is nothing, never takes place because it is never natural or artificial. It has no assignable border, no fixed perianth, no being-wreathed."[7]

The deconstructive exacerbation of indeterminacy, by making irrelevant the distinction between the natural and the artificial, evacuates the flower from the field of ontology and converts it into "nothing" within the purview of metaphysics, where it foils identity thinking and, as a result, "never takes place" as such. Defying the difference between nature and artifice, the flower is not only non-articulable but also absolutely indefinite; it misses a "fixed perianth" (the outer envelope consisting of the corolla and calyx), is distended, and does not present a recognizable form; its shapelessness, finally, is the shape of vegetal freedom in the deconstructive constellation of the event, the happening of what "never takes place." What allows the flower to quit the classical logic, built on the dichotomy of nature and artifice? Two rather formal indications, whether distilled from or inspired by Derrida's texts, will supplant at this point a full response to this question.

First, in the temporary suspension of withering away, the moment after one plucks a flower from the field and cuts it loose from the root, it is no longer distinguishable from its skillfully crafted artificial counterpart. Deconstructive sense flourishes here from the "*sans* [without] of the pure cut," unchaining the flower from its organic connection to the soil and putting it on the edge of culture as a symbol of love, religious devotion, mourning, friendship, or whatever else may motivate the culling. The culled flower is "on the edge of culture," because it does not immediately pass from a purely organic determination to the grip of "second nature." Its *Aufhebung* is arrested midway and its freedom conveys that it is "free of all adherent attachment, of all determination. Free means *detached*."[8] Exempt from conceptual determinations, this flower Derrida has picked from the margins of Kant's third *Critique* has nothing to do with knowledge, be it theoretical or practical ("The tulip is exemplary of the *sans* of the pure cut. . . . On this *sans* which is not a lack, science has nothing to say"),[9] and, thanks to this disengagement, does not submit to the conceptual discernment of the difference between nature and artifice. It is the aesthetic approach alone that will be able to hold the detached flower in this state of indeterminacy and suspension, granting it freedom from the concept, from an imperious teleology, from practical usage, and from the root—more on this later.

Second, as soon as one tries to signify or "render" the flower in an effort that, as Derrida notes, "can only fail," one is in peril of confusing this nebulous referent with the sign: "These flowers are neither artificial nor entirely natural. Why say 'flowers of rhetoric'? And what would the flower be when it becomes merely one of the 'flowers of rhetoric'?"[10] Although the flowers of rhetoric are ornate and hyper-symbolic figures of speech, they depend upon vegetal beings, supposedly as superfluous—from the vantage of conceptuality—as they themselves are. Their infringement on the boundaries demarcating the order of *phusis* from *tekhnē* is emulated, on the other side of the divide, by the conflation of the author's proper name and the common name of a flower. In the case of Jean Genet, which Derrida studies in the righthand column of *Glas*, the author becomes interchangeable with the broom-flower (in French, *genêt*): "Apparently, yielding to the Passion of Writing, Genet has made himself into a flower. While tolling the *glas* (knell), he has put into the ground, with very great pomp, but also as a flower, his proper name."[11] With this, Genet does not naturalize what is otherwise a matter of convention, his proper name; rather, he lets go of his name (as much as of himself: "yielding to the Passion"), liberates it, drops it, buries it in the ground as though it were a seed, and allows it to decompose and to rearrange itself as a flower. The trafficking of the word and the plant, the name and the flower, across the borderlines of nature and artifice subtends the version of vegetal freedom that resonates with conceptual indeterminacy.

Insightful as the deconstructive intervention may be, distention and formlessness do not exhaust the material sense of freedom, evocative of concrete shapes, figurations, and geometrical patterns, including the perfect circle in Aristotelian thought. In Hegel's *Philosophy of Nature*, too, the freest shape is the one most dissimilar from the rigidity of straight lines delimiting inorganic entities (most emblematically, a crystal); it is the flexible body, not to mention the cell, of an animal that lives up to the description of a liberated physical form. While not as stiff as a stone, the plant does not grow with the fluidity of the animal, but only gets "benumbed and rigid [*erstarrt*]" in its movement outward.[12] (The allusions to erection in this description are undeniable.) "In this respect," Hegel continues, "the plant stands midway between the crystal of the mineral sphere and the free, animal shape; for the animal organism has the oval, elliptical form, and the crystal the straight-line form of Understanding. The shape of the plant is simple. The Understanding is still dominant in the straight line of the stem, and altogether in the plant the straight line is preponderant."[13]

Remarkably, the structural position of the plant between the dead crystal and the living animal corresponds to the location *Phenomenology of Spirit* assigns to the faculty of understanding. Between "linear" sense-certainty and "elliptical" self-consciousness, this faculty still relies on an external approach to its object that confronts the subject as an alien force, similar to the plant's indifferent, external, non-mediated, not inwardly determined relation to its life. The defectiveness of mere understanding that in the pursuit of objective knowledge has not yet returned to itself is akin to the "flaw" of vegetal life bereft of self-feeling or self-relation—such is the implicit insight of Hegelian "plant-thinking." The freedom of both the plant and understanding is curtailed by their straightforwardness and simple linearity, their incapacity to become objects for themselves, as well as their naïveté. Their monumental erection fails to see itself from the outside, to register its own movement, to curve, or reflectively to circle back to itself. Bad infinity in the form of the line comes to haunt, this time, the issue of vegetal freedom.

But the architectonics of plant life, parallel to other aspects of the grandiose dialectical system, does not end there. The Hegelian plant is a biological prototype as much of understanding as of the aesthetic drive toward symmetry and regularity in the absence of freer organic forms. Despite claims made in the introduction to his lectures on aesthetics, the German philosopher is actually interested in "the beauty of nature," even if the extent of this interest is curtailed by what he sees as the true dialectical task of studying how Spirit in its own self-production takes up and transfigures external, natural beauty. So, having confessed right in the midst of his theory of aesthetics that "the plant, however, stands higher than the crystal," Hegel draws our attention to the crystal-like vegetal shape with its "chief feature" of "regularity and symmetry, as unity in self-externality."[14] (The corolla of a flower, for instance, is often formed *via* a regular and symmetrical arrangement of petals, an arrangement that is external, in that the loss of some or even all of these will not fatefully affect the entire plant.)

The transition from the vegetal world to the aesthetic sphere is all the more seamless: like the plant, which is subject to a strong influence from inorganic life, artworks presenting "unity in self-externality" occupy the systemic place of vegetal nature: "As regards regularity and symmetry, these, as a mere lifeless geometrical unity, cannot possibly exhaust the nature of a work of art, even on its external side. . . . [Rather,] the ideal work of art must . . . , even on its external side, rise above the purely symmetrical."[15] These artworks have not yet thoroughly sublated the natural world, drawing it out of its darkness and making it conscious of itself in and as Spirit, but have frozen under the spell of vegetal existence. Incapable of reflexivity and self-relatedness, they have utilized an artistic form that is pre-given, alien to them, borrowed from the plant. To learn from plants is, on this view, to unlearn the achievements of animal and human natures and dialectical self-relatedness.

In his critique of aesthetic regularity, Hegel is certainly not original, for he reiterates the lesson of Kant, according to which "all still regularity . . . is inherently repugnant to taste, in that the contemplation of it affords us no lasting entertainment."[16] The predominance of vegetal features (not of depicted plants *per se*) in a work of art infects this work with what Kant and Hegel see as the insufficiency of a life largely captivated by a dead form—"a mere lifeless geometrical unity" or "still regularity"—which, in contrast to the death borne by *Geist*, is neither mediated nor internalized. Proximity to the aesthetic ideal is henceforth measured not by an accurate representation of objective perfection and precision of symmetrical arrangements, but by the triumph of artists over the impositions of organic and inorganic nature, a criterion tantamount to enucleating the vestiges of plant life from their works. Hegelian ideality is allergic to vegetal existence, which in turn puts up material resistance to idealization. The shape of aesthetic freedom, viewed through the idealist lens, is asymmetrical: even on its "external," material, visible side, it comes through in the irregular, broken lines and imperfect ellipses that flout vegetal existence. Everything else, all symmetrical and regular arrangements, partake of the inorganic world of minerals and its reflection in the figure of the plant.

The post-metaphysical task of de-idealization finds an ally in the oppressed life of plants, typically associated with death, discarded along the path of Spirit's glorious march through the world and—thus abandoned—freed from the dialectical totality. The retrieval of vegetal existence does not intend to spark off a renaissance of the values of regularity and symmetry presumably ensconced in it, nor does it insist, in a crass inversion of Hegelian metaphysics, that mere understanding should gain an upper hand over self-consciousness. It is yet to be decided whether the values Hegel imputes to the life of plants and to what it contributes to the cultural-aesthetic sphere are themselves a product of the idealist violence that befalls vegetation at the behest of dialectics (it is enough to think of tree "limbs" to glimpse the asymmetry inherent to vegetal growth, disputing dialectical conclusions, at least empirically) and whether the faculty of understanding really finds its analog in the plant (it would be sufficient to revisit the Kantian flower that eludes understanding, a "free beauty of nature" completely isolated from the order of conceptuality and from the biological determinations of vegetal life). What becomes absolutely clear, precisely on the violent terms of the dialectical system taken to its logical conclusion, is that the plant enlivens, sets in motion, and liberates the geometrical arrangement from its own rigid confines through a unique exploration of the living character of the straight line that it embodies. How does this quasi-miraculous enlivening happen?

In the plant, we discern the cold perfection of the crystal repeated as no longer dead, as other to itself, which is why, in the words of Ponge, vegetation gives birth to living crystals, *cristaux vivants*,[17] redeeming their inorganic counterparts. Through its emulation of and close contact with inorganic entities, the plant sets free the entire realm of petrified nature, including mineral elements, if not the earth itself. When it turns mineral substances into its own means of nourishment, for instance, it extricates them from their immediate existence, dispensing to them new ontological possibilities. That is why in his philosophical treatment of the biblical Genesis, Saint Thomas Aquinas argues that rather than a superfluous adornment, vegetation is proper to the very earth it frees from the empty indeterminacy and formlessness marking the previous day of creation: "Since [the plants] are firmly fixed in the earth, their production is treated as a part of the earth's formation" (q. 59, art. 2). The scholastic circle is in this case ethically productive: the plant relieves the inorganic realm of abstract freedom by affording the earth—to which it belongs and from which, according to Aristotle, it springs as a child from its parents (*Metaphysics* 1023b6)—the chance of self-determination, while on its part the earth freely shapes and forms itself through the production of plants. The gift of material determination, offered by plants to the earth that grants them the conditions of their existence, is the indispensable supplement to conceptual indeterminacy, lauded in the deconstruction of vegetal existence.

If, as in Ponge's poetic universe, "vegetal being should be defined rather by its forms and contours [*l'être végétal veuille être défini plutôt par ses contours et par ses formes*],"[18] then the question regarding the shape of plants is a question of ontology—or in our terms, ontophytology—as opposed to morphology. Reminiscent of the Hegelian enclosure of vegetal form between straight lines and its confinement to the exteriority of a visible, geometrical shape,[19] Bergson's pronouncement that plant cell is torpid from the start—"surrounded by a membrane of cellulose, which condemns it to immobility"[20]—carries an ontological significance well in excess of the merely botanical description and consigns this shape, along with the entire being of plants, to the space of unfreedom.

The straightforwardness of plant shapes in the German and French philosophical contexts is a corollary to their lack of agility, or their ontological deceleration, one might say. Such verdicts are admittedly not without a measure of historical truth, which has nothing to do with the invention of microscopes, permitting us to visualize the immobility of the vegetal cell, and everything to do with the wholesale taming, domestication, and reification of "wild" nature stripped of the last vestiges of life. Vegetal torpor is the aftermath of civilization; it is what remains of plant life after its thorough cultivation and biotechnological transformation into a field of ruins.

If upon encountering a plant, we fail to be impressed with the exuberance of its growth and uncontrollable efflorescence, this is because its current conceptual framing is the outcome of a long history that discarded and invalidated numerous interpretative possibilities for our relation to "flora." That is why even the most comprehensive ontological investigations of the last century—those conducted by Heidegger—

force the plant into a fixed metaphysical shape, which imitates the material rigidity ascribed to vegetal cells. "Because plants and animals are lodged in their respective environments," writes Heidegger, "but are never placed freely in the clearing of being which alone is 'world,' they lack language."[21] In these lines, the physical lack of freedom morphs with great ease into freedom's metaphysical absence from the world of plants, that is to say, from their state of worldlessness, an exile from the "clearing of being" into mere "environment." And it is this very exile that in Heidegger's view determines the symptomatic dearth of language in plants and animals.

Everything is contingent, nonetheless, on the definition and the limits of language, which either conform to the traditional logocentric model, privileging the human, or exceed this model by confronting it with the material and finite sense residing in the being of vegetal beings. For what if the freedom of plants—their placement in the clearing of being through an alternative route, one that is inaccessible to human beings—is adumbrated by their self-expression, which coincides with their spatial forms? What if, in other words, these forms are indicative not of the plants' physical and ontological torpor but of their free opening unto being right within their environments?[22]

Stopping short of calling vegetal and animal lives, subject to an undeniable craving, *desirous*, Heidegger elsewhere identifies their "fixed shape" with ontological fixity, or "captivation,"[23] *Benommenheit*, typical of the being of plants and animals. Although life is inconceivable without that unrest which seems to destabilize all solid identities, its ontic movement is antithetical to the ontological stupor that, according to Heidegger, is the fate of its non-human instantiations:

> To remain thrust and forced into its own craving belongs to the essence of the living. Indeed, "the living," which we know as plants and animals, always seem to find and maintain its fixed shape precisely in this craving, whereas man can expressly elevate the living and its craving into a guiding measure. . . . If we attend only to what we need, we are yoked into the compulsive unrest of mere life. This form of life arouses the appearance of the moved and the self-moving, and therefore of the free.[24]

Heidegger's text holds in store two crucial conclusions. (1) Within the ontic-biological matrix, the craving of mere life shackles plants and animals to the compulsion of need, as opposed to human desire, which, mediated through language and conceived in a distinctly Hegelian fashion, transcends the unrest of the living in its striving toward what it does not need: the superfluous. (2) Ontologically, freedom is possible solely when the "fixed shape" of need is cast off and life with its unrest becomes a problem or a question, as soon as it is thematized and "expressly elevated . . . into a guiding measure."

It is these two conclusions, forcefully depriving the plants of their freedom, that we will submit to a close scrutiny, not only by emphasizing the play built into vegetal life, strangely indifferent to its own preservation, but also by pointing out the co-imbrication of human liberation and the freedom of plants in aesthetics, religion, and the utopian imaginary. In doing so, we will echo Ponge's pithy advice, "To free ourselves, let us free the flower."[25]

Vegetal Indifference

Heidegger's imputation of relentless needing to plants is the outcome of a powerful and to some extent necessary anthropocentric temptation to sift through and process everything that is in being from the perspective of a particular being, the human. We assume that, in analogy to our suffering from thirst and the deprivation of social contact, plants crave water and sunlight, not to mention life itself. This assumption, however, ought to be recognized for what it is—an instance of cross-species and cross-kingdom "analogical appresentation," which is a philosophical tool Husserl utilized in his *Cartesian Meditations* to solve the problem of how to describe phenomenologically the otherwise inaccessible ego of another person. (Husserl did so by conjecturing that the ego of the other is formally analogous to that of the describing subject who has no direct point of access to its contents.) As soon as human beings resort to an analogical appresentation of vegetal life, assuming that is it formally parallel to our own, the meaning of this life is presumed to be unexceptional. Indeed, it is wholly exhausted, for example, in Spinoza's seventeenth-century metaphysics, where *conatus essendi*, the attachment to existence willing its own perpetuation, explains the stubborn perseverance of all living entities, their clinging to life at any cost. The plant too is said to be under the sway of the *conatus*, a totalizing, metaphysical concept that casts life in the terms of a desire to stay alive, factored into every living being.

One of the most recent incarnations of Spinozan vitalism is Jean Grondin's *Du sens de la vie*, where, despite conceding that "the meaning of life is anterior to the human order," the Canadian philosopher claims that the plant "wants" to live: "This is also how the plant turns toward the light of the sun, since it 'wants' to live, if one may put it this way. There is, evidently, no will reflected here, but rather an aspiration of life to life [*C'est ainsi que la plante se tourne vers la lumière du soleil parce qu'elle 'veut' vivre, si l'on peut dire. Il n'y a évidemment pas ici de volonté réfléchie, mais certainement une aspiration de la vie à la vie*]."[26]

The non-conscious and unwilled "aspiration of life to life" Grondin has in mind is a contemporary replica of Spinozan *conatus* and Nietzschean will to power. This ostensibly objective and all-inclusive meaning of life is a projection onto all living beings of a historically conditioned human desire for self-preservation, a desire born from the political and economic systems that make survival ever more precarious and uncertain. Given that capitalist patterns of production and consumption prompt human subjects above all to value their own self-preservation, the plants, too, seem to partake of this desire, not the least because their survival is becoming less and less assured in the era of genetic modification and because these political-economic patterns have proven to be tremendously detrimental for the environment. The meaning of life is, on this view, one and the same for all living creatures, and plants merely supply a convenient example of the overarching logic of self-preservation.

The philosophical inquiry into the objective meaning of life neglects both the hermeneutical methodology of a patient context-specific interpretation and ontophytology, the ontology of vegetal existence, where there is precisely no intimate, inner, unified self—and therefore nothing to preserve. As we know, the plant is not one; ontologically multiple, it boasts a dispersed plurality of parts that are not internally related to each other, let alone to the individual vegetal being they comprise, and that are mutually indifferent to each other's existence, as in the case of the seed and the fruit, where "the plant has produced two organic beings which, however, are indifferent to each other and fall apart."[27] Even if a plant (for instance, milkweed) produces toxins to ward off pests or insects, it does not, strictly speaking, do so to protect itself (or better yet, its "self"). Is it then an exception within the order of the living and does this justify its sheer instrumentalization?

The involvement of vegetal beings in their world is so dissimilar from human and animal engagement that from an anthropocentric perspective it seems that plants are not at all interactive beings, uninvolved in their own existence. Vegetal indifference is the at times despised and at other times idealized counterpart to our constant immersion in affairs significant and insignificant, the existential projects that plants are unable to pursue. For a Spinozan thinker, such a life would be downright puzzling, because, at the extreme of disinterest and indifference, in the absence of the *conatus*, every thing would immediately destroy itself. But what if, contravening Spinoza's axiom, plants, as well as all other living beings, flourish only in "falling apart," in not keeping themselves intact, in not keeping themselves *as selves*? Would the idealization of vegetal freedom from the constraints of selfhood and the exigencies of self-preservation mirror a utopian image of our human freedom from necessity and need?

Across the philosophical and the literary traditions, the idealization of the indifference of plants has been rampant. It has been said that they "have no problems, no worries about the wherewithal with which to live, and, consequently, no such needs" (La Mettrie)[28] and that they "don't worry about food and lodging, they do not devour one another: no mad pursuit, no struggle to escape, no cruelties, laments, cries, words; no fret, no fever, no murders" (Ponge).[29] If so, then the meaning of vegetal existence lies elsewhere: in a life not concerned with itself, in contrast, on the one hand, to the concernful life of the animal (and the human) and, on the other, to the stone's lifeless lack of concern.[30] It is nonetheless questionable whether the indifference of plants to threats in their immediate environment is but a myth (though most likely it is!). What is certain is that plants are unperturbed and indifferent, above all to their very being, to what they properly *are*—and in this they are distinguished from human *Dasein*, with its vacillation between everyday concerns with ontic existence and the ontological care for its being.[31]

The boundary between the biologically rooted freedom from need and the ontological freedom from self has never been more porous than it is in vegetal indifference, looming as a temptation on the horizon of *Dasein*. In everyday life, human beings also usually behave in a plant-like way, *as though* they had no individuated self, *as though* they did not care for their being. They vegetate, leading what Heidegger calls an "inauthentic existence," even though, aside from exceptional limit-situations, they never quite manage to rid themselves of a concernful dispersion in the world. Their indifference is, consequently, a modification of *Dasein*'s engaged non-indifference and individuation (*Jemeinigkeit*), which falls far short of the ontological indifference of plants.

It is conceivable that human beings actually sublimate what they construe as vegetal disinterest into the corresponding ethical and aesthetic comportments, as Kant has done in the second and third *Critiques*, or as Levinas has ventured to do in *Otherwise than Being*. Below we will reassess the moral law crystallized in the categorical imperative, aesthetic pleasure uncoupled from interest, and the evacuation of essence from an ethically disinterested existent, all the while questioning what these philosophical strategies have inherited from vegetal life and how they have idealized the vestiges of this life. But for now let us pause to consider an unsettling intimacy between the philosophical idea of the gods, who are least prone to be affected from the outside, and the alleged indifference of plants.

From the diametrically opposed ends of sheer difference and self-sameness, but also material and ideal being, plants and gods present counterpoints to human (and indeed, animal) desires and involvements in the world. So much so that in *Isagoge*, Porphyry, an ancient commentator on Aristotle, brings to the fore the similarity between these two classes of beings, for whom such "negative differences" as "im-mortal" and "in-sensate" are indispensable, differentiating them from the mortal humans and sensate animate creatures, respectively (10.9–19). Roughly seven hundred years after Porphyry, Avicenna would open a similar exception in his negative definition of plants as participating in an insensate, living substance.[32] Above and below the mayhem of human life is the indifference of the gods and of the plants: the ones in flawless harmony and repose, because in the best state possible, the others in constant alteration, unrest, and metamorphosis. Human language is hardly able to convey the freedom of both these realms, which would have been ineffable were it not for the clumsy attempts at negating and at the same time maintaining mortality and sensibility in the impoverished descriptions of the super- and sub-human available to us. Still, we would do well to remember that, like that of the gods, the freedom of plants (from consciousness and from self-relatedness, from need and from the *conatus*) is not purely negative; if it appears to be so, this is due to the insufficiency of language, be it rigorously conceptual or raw and colloquial, but in any event hopelessly mired in anthropocentric references and projections.

To recap: vegetal indifference is ontological, though perhaps not ontic. When it comes to descriptive botany, this statement is admittedly less an axiom and more an indication of a general tendency of plant life, following Bergson's comprehension of the slumber of consciousness in vegetation as a dynamic tendency. The thorns of a rose and the mildly poisonous kernels of apples and peaches are no doubt some of the simplest mechanisms permitting plants to protect themselves, but these ontic facts should not detract from the ontological indifference of vegetal existence. For if we turn our attention back to a crucial aspect of plant-soul (namely, reproduction), we will find it bewildering how a flower that releases pollen or a rotting fruit that exposes its seeds delivers itself over to chance, literally throwing itself (away), potentially wasting itself. In this primal scene of dissemination, the contingency of the throw, which may have been gratuitous, "for nothing," liberates the plant from the bond of necessity first tied in Aristotelian teleology. All the while staying anchored to the earth, immersed in the immanence of the here-below, the plant selflessly "throws itself to free air . . . without hesitations or reproaches,"[33] so that the freedom of the throw, of the air, and the absence of guilt, bad conscience, or subjective depth jointly elucidate the meanings of vegetal life. It is this life's ontological indifference to being—first and foremost, to its own being—that will act as the lightning rod for a tale of two freedoms, the vegetal and the human.

A Tale of Two Freedoms

I will narrate the story about the convergence of vegetal and human freedoms from the end, which is also the closure of metaphysical history, ending in nihilism with the suspension and questioning of all hitherto accepted values.

When the mantle of sacredness and inviolability slides off from the transcendental certainties of the past, such as God, objective truth, etc., humanity is cut loose from the oppressive foundations of its existence and is set adrift, such that "we do not yet know the 'whither' toward which we are driven, once we have detached ourselves from the old soil."[34] Insofar as the topsy-turvy transcendental ground for existence—the Platonic *topos ouranios*—no longer offers us any support, insofar as the metaphysical plant has been uprooted and the physical one vindicated, insofar we have been deracinated from the "old soil" of transcendence and directed to an as yet unknown destination (if there is one), we have gained the freedom to question, to interpret, and to think, though it is by no means clear that we know how to assume this freedom. An indifferent receptivity to anything whatsoever liberates experience and thinking that, no longer loyal to metaphysical certainties, struggle to gain a new ground. This struggle, Nietzsche implies, is worth waging, unless it circles back to the "safety nets" of transcendence, completing the great Odyssey of the West. It is worth waging, in other words, if it teaches us how to live with the dangerous and terrifying freedom the nihilistic uprooting grants us, how to come to terms with the "scarecrow of the ancient philosopher: a plant removed from all soil" that we are,[35] and how to exist, to interpret, to think in the wake of all metaphysical grounds: to exist groundlessly or (what amounts to the same thing) in a way that is self-grounded, rooted in ourselves.

Groundless autonomy is in touch with vegetal heteronomy, as a result of a discursive articulation, which stresses the ground and roots of existence, and the philosophical treatment of human self-rootedness that keeps the illusion of full independence at bay. Hence the early Marx, who stands in the double shadow of Hegel and Feuerbach: "To be radical is to grasp matter at the root. But for man the root is man himself."[36] The incorrigible dialectical ambiguity of "for man the root is man himself" inculcates dependence into the strongest assertion of autonomy, so that the self-grounding of the human is at the same time her or his rootedness in another person. The successful overcoming of nihilism *à la* Marx will culminate in a non-transcendent ethics—retracing the contours of vegetal being, at once self-grounded and rooted in the other—the political name of which is "communism."

But, while it still persists, nihilism is the outgrowth of indifference, palpable in the condition of profound boredom, when human beings vegetate, careless for their being and uninvolved in the world. The Heideggerian "fundamental attunement" of profound boredom, so pervasive because not attributable to a single boring thing, is the experiential corollary to nihilism, emulating a familiar ideal type—the absolute indifference of the plant; it rarifies the field of *Dasein*'s concerns and leaves it empty "in the sense of the telling refusal of beings as a whole."[37] Thus unperturbed by the question of being and by the beings themselves, we seem to slip out of the world and to become as "worldless" as a stone or as "world-poor" as a plant or an animal. Everything, including ourselves, appears insignificant "because an *indifference* yawns at us out of all things, an indifference whose grounds we do not know."[38]

It is precisely in this emulation that human and vegetal freedoms part ways: unlike the absolute unconcern of the plant, boredom-inducing indifference is a positive modification of *Dasein*'s non-indifference, its care for being. A privation of the most intense involvement in worldly affairs, profound boredom discloses to *Dasein* its temporal constitution by slowing down and prolonging the passage of time, which lingers in and as *Dasein* after the world of things and the question of being itself have dissipated in the fog of indifference. That is why the chief ontological effect of this fundamental attunement is that upon rendering the world of beings meaningless and uninteresting, it frees *Dasein* for finding the meaning of its being in finite time.

Unwittingly feigning the "superficial" perspective of the plant, profoundly bored human beings shed age-old prejudices, not the least of which is a set of presuppositions regarding what they are *qua* human, and in so doing enact a lived, existential *Destruktion* of the metaphysical tradition.[39] We no longer know with any degree of certainty who or what we are, and in this at least we approximate the plant's non-appropriation of itself. But since it is impossible effectively to perform ontological destruction or deconstruction once and for all (since, that is, we must practice them habitually in the interminable closure of metaphysics) a necessary dose of nihilism, of the transvaluation of values, and by implication of plant-thinking will accompany human thought and praxis in the "post-nihilistic" age. The twofold liberation of the plant *in us* and of the plant *from us* is a task that never comes to fruition and is never fully discharged, one that periodically requires us to recommit to the project of freedom as open-ended as vegetal growth.

In addition to affecting nihilism and profound boredom, the reach of vegetal indifference into the ethical, aesthetic, and political spheres of human life is so extensive that the present volume will only sketch out some of its main vectors. In Kant's philosophy, saturated with calls for indifference, pure practical reason requires that before proceeding with moral judgments, the subjects purge themselves of all "pathologies," or the affectations of the will imbricated with sensibility, and act in accordance with what reason a-pathetically prescribes to all rational beings.[40] At the outset of the second *Critique* Kant postulates, as the precondition for ethics, a transcendental blindness to sensibility, to our empirical needs or desires, and with this he wishes to isolate the object of moral reason—the good—from the sensations of pleasure and displeasure. Transcendental freedom, denoting the independence of the will from empirical conditions and circumstances, is made possible by this purification, which segregates the animality of human beings from the rationality of the moral law speaking through them and which responds to the animalistic indifference of need to the voice of reason with rationality's own indifference to empirical existence in its entirety. As Kant expresses his hope, "he [=the human being] is nevertheless not so completely an animal as to be indifferent to all that reason says on its own and to use reason merely as a tool for the satisfaction of his needs as a sensible being."[41]

What has been uniformly regarded as a cold abstraction of Kant's de-contextualized approach is the secret inheritance moral philosophy receives from vegetal life, indifferent to the empirical demands of need and its satisfaction. The exaltation of the human above animal nature to the precepts of pure reason is simultaneously a descent below animality to the life of a plant, for which concerns with need and self-preservation are equally pathological, and are an exception from the general tendency of vegetal indifference. The ends of reason trump those of instinct because the object of the former is different from (and—Kant claims—superior to) that of the instinctual satisfaction of need, but it is this difference that recaptures something of the pre-animal existence and evokes a trace of the plant's freedom, around which moral law accretes. The non-transcendental condition of possibility for the transcendental sphere is the plant, which furnishes moral judgments purged of pathology and divorced from the prejudices of sensibility with an unacknowledged model antecedent to the self-legislating rational subjectivity.

Notwithstanding the philosophical abstraction and glorification of indifference, the symbiosis of human and vegetal freedoms is not celebrated, but rather concealed and disavowed. Levinas's ethical thought is complicit in this disavowal, to the extent that it resorts to disinterest as an "exit strategy" from the Spinozan logic of essence, selecting as its guide not the plant but yet another kind of transcendence, which it seeks in the non-ontological notion of God. Unlike Porphyry and Avicenna, then, Levinas forgoes the link between a theologically inspired notion of alterity and the plant. While "*esse* is *interesse*; essence is interest" and "being's interest takes dramatic form in egoisms struggling against one another,"[42] ethics signifies freedom from essence and from the "struggle of egoisms," indifference to the *conatus* tethering us to being, and, by the same token, non-indifference to the plight of the other. A shift away from the narrow egoistic perspective and from the concern with one's being takes place, for Levinas, in the wake of a "relationship with the past that has never been present" (his notion of the "trace," which ultimately reverts to God), prompting the "*investiture* of a being who is not for-itself but *for all being*" and is, therefore, "dis-interested."[43] This trace, we might add, still recoils back to vegetal beings devoid of a deeper essence and available, in the sheer exposure and exteriority of pure selflessness, "for all being." The non-evasive freedom from self at once invests the subject with meaning and responsibility, relieving ethical existence of the burden of ontological selfishness. But the source of this freedom is the plant, an instant of alterity Levinas does not thematize and would most likely not hail as such.

As in Kant, the highest level of transcendence in Levinas merges with and overlays the lowest, if one is willing to entertain the suggestion that Levinas's text stages a masquerade of the plant, which is disguised as the non-ontological divinity and synonymous with the relation to the other. To be sure, he no longer casts the indifferent egoism of need, or—in keeping with the prosaic metaphor of *Totality and Infinity*—a hungry stomach "without ears,"[44] in terms of the animalistic dimension of human nature. Rather, Levinas rids this egoism of all biological and anthropological overtones as a consequence of the ontological interpretation that takes Spinoza at his word and recognizes in the depths of ontology an attachment to being, the passionate interestedness in a being's preservation and perpetuation in being.

The choice Levinas makes available to us is not between the indifference of reason and pathological sensibility, but between ontological egoism, shored up by reason and sensibility alike, and disinterestedness, transporting ethical subjects "beyond essence," while also granting to them the possibility of existence "otherwise than being." If it is fully independent from reason, the freedom *from* essence and *from* ontology cannot enter the content of conscious representation, though it bestows meaning upon representations from the exterior position of a trace. Non-representable thinking, an-archically preexisting ontological thought, the thinking that is but a trace of thinking and that, unaffected by the *conatus* or by being-for-self, puts *logos* as a whole under erasure, all the while breathing sense into it—this is none other than plant-thinking, which Levinas treats under the rubrics of non-ontological religion (i.e., ethics) and the "saying without the said." To approach the non-egoistic indifference of the "otherwise than being," it may be enough to summon whatever remains of the plant in us, without resuscitating religion at the dusk of metaphysics.

Kantian and post-Kantian philosophies of aesthetics also give prominence to the non-representable facets of vegetal life that do not fall under the purview of essence and ontology. In *Critique of Judgement*, a flower is famously deemed to be the "free beauty" (*pulchritudo vaga*) of nature, free because extracted from the Aristotelian logic of finality, unconditioned either by the concept of beauty or by considerations of practical utility that would be subject to particular ends.[45] The non-conceptuality of free beauty, incomprehensible within the framework of the first *Critique* and opposed to the functionally or conceptually contingent "dependent beauty" (*pulchritudo adhaerens*), does not postulate standards against which actual flowers could be judged. There is no transcendental ideal of a beautiful flower, for floral beauty is grounded solely in itself, in each singular flower. It arises freely, *sui generis*, puts up a powerful resistance to idealization, and thus offers an aesthetic corrective to the pure-theoretical and moral-philosophical excesses of normativity. And in this it is akin to the being of plants themselves.

As the Latin word *vaga* in *pulchritudo vaga* hints, the freedom of a beautiful flower, viewed from the standpoint of conceptual thinking, borders on vagueness and indeterminacy, a strand of Kantian aesthetics, which Derrida's deconstruction enthusiastically adopted. "Where an ideal is to have place among the grounds upon which any estimate is formed," Kant writes, " . . . there must lie some idea of reason according to determinate concepts, by which the end underlying the internal possibility of the object is determined *a priori*. An ideal of beautiful flowers . . . is unthinkable."[46] In the jargon of conceptuality, these flowers are meaningless, represent nothing, do not point beyond themselves, and say nothing to us, least of all about their belonging to the worlds of *phusis* or *tekhnē* or both. Which is why, in a single breath, without making a transition from nature to artifice, Kant cites, as an example of free beauty, a flower growing in the field alongside "foliage for frameworks or on wallpaper."[47] The beauty of a flower is indifferent to its "natural" or "artificial" provenance, but if we cannot think the floral object based on the possibility of determining it *a priori* (for instance, as something that "exists according to nature"), then this flower deserts the Kantian subject–object opposition and above all the objective position that would have been allotted to it. Beautiful flowers are emphatically not objects, but neither are they the ungraspable things-in-themselves demarcating the internal limit or frame of the first *Critique*. They must be thought otherwise, without the mediations of conceptuality and along the general lines of plant-thinking, to which Kant's *Critique of Judgment* makes a substantial contribution.

Deconstruction has already accepted the challenge of extra-conceptual thinking and its relation to the concept above all in the investigations of the positive meaning locked in the frame's (more generally, parergon's) insignificance and non-signifyingness. The point, however, is to re-engage the plant, among the other tropes of "nature," in the thinking that derives from it. The absence of a conceptually mediated meaning does not signal the voiding of sense in the flower that represents nothing, but conversely announces a shift in the directionality of sense, whereby, as Nietzsche put it, "having been released from this struggle [for existence] by a stroke of good fortune," the flower "suddenly looks at us with the eye of beauty."[48] While in many respects he anticipates the conclusions of deconstruction *avant la lettre*, Nietzsche is entirely Kantian here, linking beauty as such and, in the first place, floral beauty to the liberation of the flower from the realm of necessity and from the considerations of utility. The beautiful flower ceases to be an object of human regard, instead looking at us with the de-subjectivated and impersonal "eye of beauty" because we do not exactly need it. Exempt from transcendental determinations, it excites, as though by contagion, in those who behold it, a certain spontaneous and unexplained enjoyment, culminating in subjective freedom. Plant liberation is indispensable for the possibility of human liberation. How so?

On the ruins of the subject–object opposition, the liberation of the flower from transcendental constraints triggers the emancipation of those who come into contact with it. This is so not only because, upon encountering this non-object, the faculty of imagination is unencumbered by conceptual restrictions and by what, for understanding, appears to be "impossible," but also because flowers "please [*gefallen*] freely and on their own accord."[49] The pleasure they awaken in us falls outside utilitarian calculus, in that, withholding from us both intrinsic and extrinsic benefits, it is, like the plants themselves, disconnected from the logic of ends.

It is in this light that we ought to reassess the thesis concerning the disinterestedness of aesthetic pleasure: the indifferent judgment of pure taste, oblivious to the real existence of its object, reproduces the indifference of vegetal life, unconcerned with its being. If "everyone must allow that a judgment on the beautiful which is tinged with the slightest interest, is very partial and not a pure judgment of taste,"[50] then the non-transcendental purity of such judgment is achievable by means of an apprenticeship in vegetal freedom, a vicarious learning from the plants that should enable us to live, extra-conceptually and extra-morally, with a modicum of aesthetic-vegetal vagueness. The attitude of disinterested pleasure, antecedent to aesthetic judgments, requires careful cultivation, taking its cues from the flowers' spurning of idealization.

The other sense of *pulchritudo vaga*, "an itinerant beauty," evokes the sort of errancy we least of all associate with vegetal life. After all, how could plants be itinerant if their roots are firmly anchored in the earth, making them, in contrast to us—"their vagrant kinsmen"—not "superfluous adjuncts to the world, intruders on the earth" but agents of the earth's self-determination?[51] Derrida's *Truth in Painting* contains some of the most perspicacious reflections on the plurivocity of *pulchritudo vaga* in Kantian aesthetics and on the implications of non-conceptual beauty, which, rather than marking the plant in general and in the abstract, pertains to a singular errant flower, for example, the tulip. Disengaged from the ideal teleology of the concept (its truth), as much as from the material finality of the reproductive cycle, the "tulip is beautiful when cut off from fecundation. Not sterile: sterility is still determined from the end, or as the end of the end. . . . The seed loses itself, but not . . . in order to be lost or to refinalize its loss by regulating the diversion according to turn or return, but otherwise. The seed wanders [*s'erre*]."[52] When the seed loses itself, it betrays in various ways the generality of the species it is expected to represent. On the one hand, the "lost" seed does not germinate, dodging its reproductive functions; on the other hand, as in the case of an apple seed, it gives rise to a new plant, which has little in common with its progenitor,[53] and so institutes each time an altogether different species limited to but one tree, wherein the singular and the universal coincide. The wandering of the seed rehabilitates chance (*tukhē*), which Aristotle had banished from the regularized order of *phusis*, and portends the beauty of what is unpredictable, unknown, discrete, discontinuous, and non-reproducible even in the middle of the reproductive cycle.

The Kantian tulip gains its freedom when, capitalizing on the internal rupture and the discontinuities within the "natural" cycle, it is extracted from the universality of self-reproducing life, from the chain of vegetal metamorphoses, and from the genus under which it would have been subsumed. The actual flower ceases to represent a mere transitional stage, at which—analogous to the plant as a whole, supposedly representing an intermediary step between inorganic nature and the animal—it points beyond itself to a higher end (e.g., the fruit) or to the species to which it belongs. On the brink of non-being, passing or withering away, it begins to signify nothing but itself.

It would thus be inaccurate to claim that the act of culling definitively transplants the tulip into the realm of death, the mere privation of life ("the end of the end") symbolized by sterility. Privation is the work of capitalist expropriation, with the entire economic system coming to life on account of dead human labor and non-renewable vegetal growth, incredibly productive of value. Economically exploited by corporations that genetically engineer and patent seeds, made to yield sterile crops so as to force farmers to buy more seeds the following year, the calculated loss of the plants' finality is the "diversion according to [the] turn or return" of capital. But the inherent errancy of the seed and the "wandering" of the flower take part in the kind of aesthetic play that, rather than initiating a mere detour on the path to the final end, stands for their insubordination to the idea of finality. Only in becoming superfluous, unproductive and un-reproductive, is the tulip beautiful.

It is hardly surprising, then, that the quintessential category of post-Kantian aesthetic philosophy, namely play, liberating beings from the realm of necessity and from the cognate values of efficiency and productivism, also applies to non-human nature, including plants. Play excuses one from the demands of reason and the concerns for one's self-preservation (which in the end may amount to one and the same thing) because it is, to begin with, in effect in "mindless Nature" (*vernunftlosen Natur*), luxuriously expending its powers (*Kräfte*) in a non-productive manner. Schiller's memorable example of this aneconomic expenditure, which earns the appellation of "physical play" (*physische Spiel*), is precisely a tree: "The tree puts forth innumerable buds which perish without developing, and stretches out for nourishment many more roots, branches, and leaves than are used for the maintenance of itself and its species. What the tree returns from its lavish profusion unused and unenjoyed to the kingdom of the elements, the living creature may squander in joyous movements."[54] Since a "superfluous" bud might not open, might not give rise to a flower, its freedom implies that it may not arrive at the teleological destination prescribed to it. If the plant could enjoy its existence, the non-productive buds and the unnecessary—within the economy of nourishment and procreation—roots, branches, flowers, and leaves would have furnished the objective evidence of its *jouissance*. In its "lavish profusion" (*verschwenderischen Fülle*), the tree approximates the enjoyment of the animal, surpassing the enclosure of need to which Heidegger will later confine all living non-human beings.

Deviation from the productivity of nature effects a break in its reproductive machinery; the plant no longer fits the rigid parameters of the vegetal soul, responsible for the activities of nourishment and reproduction, and so abandons its ownmost mode of being, acquiring the freedom to be otherwise than it is. Similar to the Kantian tulip, the luxuriating parts of the tree are cut off from fecundation not because they are sterile or detached from the root but, on the contrary, because they abound in possibilities irreducible to their reproductive potency, even if such possibilities might be put at the behest of ends as yet unknown to the human observer. Ignorant of the demands of self-preservation and the preservation of the species, the tree's playfulness confounds the teleological account of nature, to the extent that it literally regresses to the "kingdom of elements" (*Elementarreich*), dispenses its unused parts back to the inorganic world, and disregards the highest ends of fructification.

In his text Schiller rehashes the medieval and early modern notion of *natura naturans* ("nature naturing," nature conceived in an active sense—as what we might term *natura ludens*) "nature playing," wasting itself, not living up to its potential but reveling in a profusion of non-realizable possibilities. Still, Schiller complicates the tale of the two freedoms when he abruptly interrupts the interplay between physical and aesthetic play, contending that a tremendous leap (*Sprung*) must be taken in order to reach the aesthetic permutation of such useless activity. What this leap involves (and what is missing from the physical play of nature) is the appreciation of "free form," *freien Form*, through the faculty of imagination.[55] The modern overvaluation of formal freedom comes at the expense of its material (or "physical") variety, relegated to the lower tier of the superfluous within the two-tier Schillerian system. The freedom of the plant is the ground to be left behind in the human aspiration toward self-realization that places its bets on the independent creative power of the subject.

Schiller's *prima facie* dualistic construction of freedom, split between two kinds of superfluity, reveals, on a closer reading, a mediating stratum that transforms the so-called leap into a mere step. In the development of its genotype and phenotype, human imagination first indulges in "material play, in which, without any reference to shape, it simply delights in its absolute and unfettered power."[56] The "*free sequence of images*" present to our imagination is the inheritance humans have received from the material play and innocent enjoyment prevalent in vegetal and animal lives. Lacking the principle of formal organization in itself, it is at the mercy of what is posited outside the subject, who sensuously receives such externally produced impressions. The material freedom of imagination is the echo of vegetal freedom in human beings, but so is the formal aesthetic play-drive, indifferent to the real existence of its object. To let the plant in us flourish, to give free reign to imagination in its materiality, we should forget the formality of "high culture," which corresponds to the upper tier of play, and to abandon ourselves to what Schiller decries as crude taste: "first seizing on what is new and startling, gaudy, fantastic and bizarre, what is violent and wild."[57] Nietzsche's Dionysian art, itself linked to the intoxicating power of a plant (the fermented grape), is no doubt crucial to this appeal, as is Deleuze and Guattari's take on "drunkenness as a triumphant irruption of the plant in us."[58]

At the same time we must be on our guard against the all-too-prevalent idealist illusion that vegetal life is the realm of purity and innocence. The plant's ontological indifference and lack of concern bespeak its freedom from conscience, but it is an anthropomorphic projection alone that codifies these qualities, as well as everything connected to play, in terms of innocence and lightheartedness. Hegel spearheads the problematic codification, both insofar as he considers plant-being on the whole to be indifferent to sexual difference and insofar as he juxtaposes the "innocence of the *flower religions* [*Unschuld der* Blumenreligion], which is merely the self-less idea of self" to "the earnestness of warring life [. . . and] the guilt of *animal religions*,"[59] turning the human relation to plant life into a transitory and ephemeral idyll of peace and tranquility, which had to be negated by the sacralization of bloody animal offerings as soon as Spirit was capable of relating to itself. The price for the peacefulness of flower religions is the "quiet [*Ruhe*] and impotence [*Ohnmacht*] of contemplative individuality" engrossed in the selflessness of vegetal life,[60] where to be absolutely innocent is to be weak in the ontological sense of weakness, that is to say, to lack the boundaries delimiting subjectivity and cordoning it off from the otherness of nature. Borrowing some of the features from the centerpiece of their worship, the practitioners of flower religions merge with the world and experience the most overwhelming "oceanic feeling," since, again like the plants they deify, they have not yet mustered enough subjective rigidity to assert themselves against this world. The orientalism of nineteenth-century philosophy, which associates the cult of flowers with "the East," coincides with its denigration of plant life.

Adorno and Horkheimer, in *The Dialectic of Enlightenment*, are thoroughly Hegelian in this respect: interpreting the story of the Lotus-eaters related by Homer, they read it along the lines of the stupid innocence of flower religions. It is said that lotus flowers, when eaten, induce the idyll of forgetfulness that nullifies self-preserving reason, but Adorno and Horkheimer note that "it is only an illusion of bliss, a dull aimless vegetating," which at best "would be an absence of the awareness of unhappiness."[61] Critical theorists paint the Homeric image as an irresponsible utopia, capricious and immature, ecstatically oblivious to the realm of necessity—foreclosed to the plant—and therefore forgetful of the rationale behind work, meant to satisfy our needs. Those who eat flowers disregard the nourishing potential of the fruit, let alone of animal flesh; the Lotus-eaters, as well as the Homeric Cyclopes, whom Odysseus encounters on the next stop in his voyage and who plant nothing yet enjoy the bounty of vegetal nature,[62] trust the plants, surreptitiously learn from them, and subsist on them without putting any effort into their cultivation. To proponents of the Enlightenment, this trust is barbaric and unacceptable.

Despite the seeming anachronism of the Homeric image and its critical reading, at issue here are today's most pressing questions of political and economic freedom: How do we put an end to alienated labor? What kind of political-economic framework is necessary for catering to the needs of all? Which forms should human emancipation take, if it is to release its beneficiaries from the multilayered straightjackets of calculative reason, the concerns with self-preservation, and the capitalist harnessing of need and of labor? And what is the role of nature (specifically, of vegetal nature) in the struggle for emancipation?

To the extent that it does not abandon the dominating attitude to the world of plants, critical theory, with the possible exception of the thought of Herbert Marcuse, is not critical enough. Such theory meets its limit in straining to convince its adherents that the innocence of plants is unsustainable in the social, political, and economic arenas and that we are duped by plants that, instead of freeing human beings, compound our misery, merely anaesthetizing us to the effects of scarcity and robbing our existence of its proper goals or ends. Under their inebriating, narcotic influence, drunk with the illusion of innocent freedom, we forget the Freudian reality principle and single-mindedly obey the pleasure principle, abdicating all responsibility not only for ourselves but also for our contemporaries and the future generations of those who will be unable to enjoy the ephemeral fruit of the untenable utopia of the Lotus-eaters.

Much in these critical theses is nonetheless contingent upon the uncritical acceptance of the idealist hypothesis that postulates the innocence of vegetal life, treating the question of the plant in the spirit of "verginity," as Derrida names it,[63] alluding at once to the virginal status and the being-on-the-verge of the plant that is but a vanishing mediator of natural religion. The deconstructive counterthesis, set against the idealist hypothesis, is that the flower is *coupable*—in the Leavey and Rand translation of *Glas*: "cuttable-culpable"—non-innocent, always already entangled in phallic imagery: "The phallic flower is cuttable-culpable. It is cut [*se coupe*], castrated, guillotined, decollated, unglued."[64] Containing the sexual organs of the plant, the culled flower both assuages and exacerbates castration anxiety: *assuages*, because the knife spares the man's sexual organ and slits, in a sort of sacrificial ritual, a non-human being that supplants man; *exacerbates*, because, despite (or better, due to) symbolizing romantic love, it is an ever-present reminder of the possibility of castration and death.

The argument against the innocence of the flower goes right to the heart of idealism and announces the return of the repressed vegetal sexuality that had to be banished, or rendered superfluous, by Hegel in order to refashion the plant into the symbol of purity, set apart from the fallenness of animal carnality. The emancipation of the flower will not come to pass without plant-thinking piercing through layer after layer of the idealist repression weighing upon it.

On the side of the human devotees of flower cults too, innocence and naïveté are questionable. The appreciation of flowers and of beauty testifies to the achievement of a certain degree of freedom for those who can spare some of their time for aesthetic contemplation, just as select members of ancient Greek aristocracy could entertain philosophical concerns once their immediate material needs were satisfied by the slaves. Otherwise, in the absence of universal economic freedom, the flower simply cannot be extracted from the considerations of utility, as in African cultures, where "flowers play almost no part in religious observance or everyday social ritual" and where they are mentioned only "with an eye to the promise of the fruit rather than the thing itself."[65] Luxurious, excessive, superfluous, they are complicit in the injustice lying at the foundations of limited freedom, accessible to very few at the expense of the many and regulating the degrees of attention one will pay to these "free beauties of nature."

While it is true that the emancipation of human beings is incomplete without the liberation of vegetal life, plants will not be free unless the political and economic conditions responsible both for their oppression and for our appreciation of them change as well. But if critical theory has failed in the difficult task of plant liberation, then what can we expect from Marxism, from which this theory draws some of its inspiration? In Marx's early writings, religion—the "opiate of the people," invoking a drug, incidentally, derived from the poppy flower—is metaphorically equivalent to the "imaginary flowers" adorning and concealing the chains of economic exploitation. Somewhat predictably, Marx puts his faith in the distinction between the real and the ideal, the living flower and its imaginary counterpart, as well as, within the scope of epistemology, science and ideology. Ideology critique, Marx argues, will be hostage to the very chimeras its object tirelessly spawns, if it stops at ideality, to which it is negatively related, and does not constructively engage with the Real. But, even after taking this further step, the liberated "man" will not let plants be, will not leave them alone: "Criticism has plucked the imaginary flowers from the chain, not so that man shall never bear the chain without fantasy and consolation, but so that he shall cast off the chain and gather the living flower."[66]

Following a venerable tradition of German philosophy, human freedom from necessity culminates in the plant's detachment from its organic connection to the earth: such is the only vegetal liberation ideology critique is capable of envisioning. That the power of Hegelian and post-Hegelian criticism alike is insufficient to break the chains of oppression becomes evident in its uncritical endorsement of the act of "gathering the living flower," symbolically invested with a sense of utopian freedom and with a disregard for the value of vegetal life. What is required in its stead is not a more powerful criticism but an infinite loosening up, a weakening of the self's boundaries, commensurate with the powerlessness (*Ohnmacht*) of the plants themselves.

5. The Wisdom of Plants

If one has no belief of anything, but is equally thinking and not thinking, how would one differ from a plant?
—Aristotle, *Metaphysics*

Flora enseigne ces vérités à qui sait l'écouter, puis l'entendre.
—Michel Onfray, *Les formes du temps*[1]

The reflections on vegetal intelligence gathered in this chapter ought to be taken as a footnote to Nietzsche's provocative suggestion that, on the quest for the "principles of a new evaluation" (the title of book III of *The Will to Power*), "one should start with the 'sagacity' of plants."[2] The re-valuation, from the ground up, of all values after the de-centering of the human and in the aftermath of the ensuing nihilistic malaise requires a thorough demolition of beliefs about the nature of "knowledge" and "truth" held dear even by the sworn enemies of the dogmatic slumber of reason: Descartes, Kant, and Husserl. It entails a wholesale rejection of claims to the objective validity of knowledge and at the same time the unmooring of the conditions of possibility for knowing from human subjectivity or, on the material-physiological level, from the brain and the central nervous system.

The epistemology that will emerge from this difficult exercise will not stop at supplanting epistemic anthropocentrism with a philosophical zoomorphism, no matter how justified the inclusion of animal instincts under the purview of knowledge may be; it will also, and more radically, revisit the Parmenidian postulate that "to be and to think are the same," which in the case of vegetal life will mean that plant-being and plant-thinking are the same. Admittedly, as we have come to realize, ontophytology overflows the strict confines of traditional ontology, in that vegetal existence belongs to the realm of the "otherwise than being." But in the spirit of the Parmenidian *adequatio*, there is no reason why this excess should not be reflected in a corresponding epistemophytology—the epistemology of plants—that similarly oversteps the bounds of the conventional theories of knowledge. Before making a start with the "sagacity" of plants, one must reckon with their mode of being (that is to say, with the multiple meanings of their life), transposing the functions of the vegetal soul onto the discourse of thinking.

If plant-thinking derives from vegetal being, then the preceding chapters, with their focus on non-metaphysical botanical ontology, will have yielded intimations of the plants' "sagacity." What I have earlier referred to as the "non-conscious intentionality of vegetal life" amounts to an essentialism-free way of thinking that is fluid, receptive, dispersed, non-oppositional, non-representational, immanent, and material-practical, provided that each of these descriptors be first extracted from the metaphysical context of its origination and ongoing usage in a procedure that will dissociate it from the oppositional relations underwriting and at the same time devaluing its sense. As a consequence of formulating the tenets of epistemophytology, we will undergo an apprenticeship in a post-metaphysical way of thinking and, in the same stroke, get a better grasp of the suppressed vegetal sources of human thought, which is both an idealizing and an idealized permutation of plant-thinking.

Non-Conscious Intentionality

"Non-conscious intentionality" inverts the Levinasian notion of "non-intentional consciousness," the concept that encapsulates much of the immanent critique the French philosopher directed against Husserl's phenomenology.

At first blush, the term Levinas has introduced into phenomenological vocabulary appears to be an oxymoron, given that for Husserl, intentionality is precisely the being *of* consciousness, its directedness-toward something outside of itself in a constant process of self-transcendence that, in thematizing itself, in becoming conscious of itself, never leaves itself behind (this is the source of Husserl's theoreticism). Non-intentional consciousness, on the other hand, would be one that lacks directionality, and so it would not be a consciousness, let alone self-consciousness, at all. The seemingly impossible "reduction" of intentionality nevertheless suits Levinas's philosophical project, to the extent that it undoes the ontological and totalizing construal of the human and affords us access to alterity, the ethical realm "otherwise than being" that predates ontology itself. "It is not illegitimate," Levinas defensively notes, "to ask ourselves whether, beneath the gaze of reflective consciousness understood as self-consciousness, the non-intentional, lived contrapuntally to the intentional, retains and renders up its true meaning."[3]

Above all, the non-intentional is not directed to itself, eschewing the reflux movement of all conscious and critical-theoretical activity that attends to itself while attending to the other. Something of this non-intentionality is present in the plant, which boasts neither a self to which it could return, nor a fixed, determinate goal or purpose that it should fulfill. Although not synonymous with the collapse of meaning, the breakdown of intentionality is a harbinger of the dissolution of the Aristotelian teleology that governed everything Husserl had to say on the subject of the relation between the intending (*noesis*) and the intended (*noema*). Instead of pursuing a single target, non-intentional consciousness uncontrollably splits and spills out of itself, tending in various directions at once, but always excessively striving toward the other. The plant, on its part, is a living attestation to the crisis of teleology and to the exuberant excess of the living and its meanings, which accords with and perhaps feeds, without ever satisfying, the ethical excess.

Regardless of all the resemblances between the two, the plant's non-intentionality crucially differs from that of ethical subjectivity. Rather than furnishing a true image of transcendence, the uncontrollable tending of vegetal life corresponds to George Bataille's depiction of animal experience as pure immanence (the animal moves like "water in water"),[4] as immersion in its milieu, with which it fuses. It would be more accurate, consequently, to conceive of plant-thinking in terms of a "non-conscious intentionality," where meanings proliferate without the intervention of conscious representations. In what ways, then, is vegetal intentionality "non-conscious"? And what gives us the right, despite everything, to designate it as "intentional"?

If intentionality does not belong exclusively to consciousness, one could conjecture that it equally pertains to vitality itself, to the contingent itineraries and detours life takes in its active unfolding, or—if one were to resort to ancient Greek philosophy—to the vegetal soul, which not only unites in itself the reproductive and the nourishing capacities but also subsequently engenders the other psychic strata, such as the sensorium. Since life and consciousness are subsets of invention or creative activity,[5] the non-conscious life of plants is a kind of "thinking before thinking," an inventiveness independent from instinctual adaptation and from formal intelligence alike.

Consciousness appears to be a puzzling exception when it is judged against the backdrop of the sheer nullity of consciousness (*conscience* nulle) peculiar, for instance, to the stone, not when it is contemplated in the context of the relative non-consciousness of a plant, in which, as Bergson notes, "consciousness is nullified" (*conscience* annulée).[6] A consciousness nullified (literally, "annulled") maintains the possibility of a sudden awakening, of passing from a dormant potentiality into an actional mode. But it does not need to connote an epistemological flaw, a deficiency that would be remedied if only plants were to make an evolutionary transition, in Bergsonian terms, from material knowledge to the formal knowledge of intelligence. Rather, it should be studied on its own terms, forgoing teleological references to the "higher order" apparatuses of knowing that presumably distinguish animals and human beings. Only in refusing to treat intelligence as an exception in the order of life and in the evolutionary process will we gain admission into the yet-uncharted terrain of plant-thinking.

Just as psychoanalysis confirms that memory-fragments are often unavailable to the human psyche in the shape of conscious representations due to the fact that traces of the situations of trauma and extreme repression are imprinted directly on the unconscious, so plant-thinking attests to the existence of a non-conscious, involuntary memory in plants. To say that vegetal beings possess memory is to claim that they have a past, which they bear in their extended being and which they may access at any given moment, or more simply, it is to assert that they are temporal beings through and through. Their memory is, in Nietzsche's estimation, imageless and non-representational: "E.g. in the mimosa we find memory, but no consciousness. Memory of course involves no *image* in the plant. . . . Memory has nothing to do with nerves or brain. It is a primal quality."[7] Vegetal memory arises at the site of material inscription on the body of the plant and contributes to the register of physical stimuli (touch, exposure to light or darkness, etc.) that, having already affected the plant, may be retrieved after a delay, when the actual stimulus is no longer present.

Contemporary cell and molecular biology abounds in examples of "information retrieval" by plants; it will suffice to mention two of the most emblematic. Barley leaves unroll if they are exposed to red light, as long as they contain calcium. If, however, calcium is removed from the plant at the moment of exposure and added up to four hours after the exposure, the same effect (the unrolling of the leaves) takes place without the reintroduction of red light.[8] The plantlets of flax respond to various stressful stimuli, such as drought, wind, or even physical manipulation, with a depletion of calcium from their cells in a process that takes approximately one day. Yet it was found that the morphogenic signal regulating calcium levels does not diminish in intensity for up to eight days after the end of the "traumatic" event.[9]

These examples demonstrate that what Nietzsche chanced upon in his reflection on the mimosa—the sensitive plant *par excellence*, one that closes its leaves in response to touch or absence of light—is in fact a more general tendency of vegetal beings to store imageless and non-representational material memories in their cells, and so to retain a trace of the remembered thing itself, in place of its idealized projection. Whereas humans remember whatever has phenomenally appeared in the light, plants keep the memory of light itself. Conceived in a non-anthropocentric fashion as a "primal quality," memory, inherent in plants at the cellular and molecular levels, comes to describe any network of traces, of which consciousness is a highly circumscribed instance. It is the very fact or facticity of impression, of an imprint, or better, an ex-print, that forms the register of what a living being has undergone in its lifetime.

Non-conscious memory is but one constituent of the vibrant and multidimensional intelligence of plants, which falls under the rubric of what Schelling, in his *First Outline*, calls "sensibility," or the "*universal* cause of life" that, in his words, "must also belong to plants."[10] Schelling believes that sensibility is not only the cause of life but also, along with its opposite (irritability), the quantum of force permeating every living entity. It is therefore possible, following his hypothesis, to map living intelligence, if not the intelligence *of* life, on the ever-shifting continuum of sensibility–irritability.

While sensibility in plants approaches zero-degree, the minimum of irritability in them ensures their survival and endows them with a certain non-conscious thinking: "Magnetism is as universal in universal Nature as sensibility is in organic nature, which also belongs to plants. . . . Sensibility [in them] passes into irritability. . . . Where the higher factor of sensibility (the brain) gradually disappears and the lower gradually attains preponderance, sensibility also begins to fade into irritability."[11] If irritability defines a passive and non-conscious thinking, then to live is already to think, and the life of plants is co-extensive with the mode of thinking appropriate to them. The brain and the central nervous system do not invent a new function but offer a novel solution to the old problem of life, which had been already raised, differently, in the very ontology of plants.

It is fair to say that Schelling is not alone in postulating continuity between life and thought; in the history of philosophy, Aristotle, Plotinus, and Spinoza are among his illustrious predecessors, emphasizing the immanence of thinking in living.[12] When Aristotle ponders the relation between the three kinds of soul, he insists that "the earlier type always exists potentially [*dunamei*] in that which follows" (414b29–31), implying that the vegetal soul carries on "potential" existence in the sensitive psyche of the animal and in the rational soul of the human. The human is not only human but also a potential animal and a potential vegetal being, though exactly how these potentialities make themselves known in us is a separate question, to be considered below.

Plotinus adheres to the Platonic view, discussed in earlier chapters, according to which there is a faculty of desire in the vegetal soul of plants, though he specifies that no living being becomes conscious of its desire until the desiring impulse reaches the sensitive faculty (*Enneads* 4.4.20.14–17). The obscure desire of plants is distinct from the knowledge afforded by sensations and processed as judgments only inasmuch as these faculties do not yet consciously register it.

Spinoza's immanentism invalidates the absolute (i.e., substantive) difference between what thinks and what does not think, in that it teases out the unity of thought present in different degrees of clarity in the affects and in their conscious representations. However questionable, the ontological "power" of the plant (the power of the *conatus*, what keeps it fast to being) is a non-conscious, not-yet-clarified thinking, which becomes active in different modes and to a greater degree in animals and human beings. Being and thinking are thus united in their pursuit of survival.

What these diverse philosophical approaches have in common is that, in establishing an identity between living and thinking, they internally homogenize both spheres, turning all differences within them into so many folds in the immanent fabric of Life or Thought and investing non-conscious existence with a teleological anticipation of consciousness that promises to elevate and clarify its muddled impulses. But the problem (or the blessing, depending on how one approaches it) is that life oft-times does not comprehend life: a form of life does not always communicate explicitly with another such form, for not all living beings share the same set of concerns, life-worlds, and modes of signification.[13] Where radical disconnects are apparent on the continuum of cross-species and cross-kingdoms intelligence, we find ourselves grappling once again, albeit this time on the epistemological or epistemophytological front, with the problem of the "commonality" of plant-soul that, already in antiquity, has become indistinguishable from the sameness of life's cause. Conversely, reclaiming the differences and nuances of plant-thinking requires counterbalancing the theoretical attention we pay to the non-conscious with a focus on the specific intentionality of vegetal life.

The non-conscious *intentionality* of plants finds two outlets, which jibe with the capacities of the vegetal soul to seek nourishment and to reproduce itself. The turning and striving of a plant toward the sun is perhaps the most iconic illustration of its non-conscious *noesis*, or act of intending, which, in the words of Gustav Fechner, supplies the evidence of the plants' "soul-life" (*Seelenleben der Pflanzen*),[14] animating vegetal bodies. Thus, citing potatoes sprouting in the cellar, Hegel marvels at how the sprouts "climb up the wall *as if they knew the way*, in order to reach the opening where they could enjoy the light."[15]

But what is even more remarkable, first and foremost in the Hegelian philosophy of nature *proper*, is that the intentionality of nourishment parallels the intentionality of perception, willing, judging, etc., *as if*, along with these exemplary processes, it were a modality of knowledge, "*as if* [als ob] *they knew the way*." The theoretical fiction of *als ob* brings home the classical phenomenological point that across the spectrum of intentionalities, the intended singularizes the intending: the consciousness—as much as the non-consciousness—of something becomes itself thanks to that of which it is conscious (or not conscious). "It is from light that plants first get their sap," Hegel states, "and in general, a vigorous individualization; without light they can, indeed, grow bigger, but they remain without taste, color, and smell."[16] The growing acquires both its quantitative and its qualitative determinations from that toward which it grows, i.e., from light, a non-ideal *noema*, unavailable for the acts of appropriation and domination. Similarly, the judging consciousness is convoked by that which is judged, the willing by the willed, and so on. In German panpsychism as much as in dialectics, non-conscious vegetal striving toward the sun is the prototype of conscious life.

But the analogy also has its inherent limits. The intentionality of the plant is not unidirectional, given that the roots, too, seek nutrients, navigating a veritable environmental maze, sensing humidity gradients of the soil,[17] and avoiding movement in the direction of other nearby roots.[18] A combination of passive growth and what appears to be an active "foraging" for resources positions this intentionality on the hither side of the distinction between passivity and activity. Plant-thinking neither grasps its object—it has none!—nor impassively freezes in sheer inaction but instead operates by the multiplication of extensions, by contiguity with and by a meticulously adumbrated exposure to that which is materially thought in it. It matters little that vegetal life does not objectify what it strives toward, or that it "is related to light as well, but . . . is not open to light *as* light,"[19] because it does not at the same time relate to itself. *Contra* Heidegger, the plant has a world (if not worlds) of its own, if in this "having" we manage to discern the overtones of a non-appropriative relation to the environment, *with*, *in*, and *as* which vegetal beings grow.

If dynamic *ex*tension is at the core of vegetal *in*tentionality (growth being understood as extended intentionality), then recent philosophies of the body should resonate with plant-thinking. And indeed, Maurice Merleau-Ponty's pre-reflective intentionality of corporeity, or "the language of the body," shares various features with that of the plants. Located in a determinate context, the body exhibits a non-conscious intentionality in its very motility—for instance, in the minute movements of muscles, restricted to the peripheral nervous system, that make up the act of raising one's hand. For the corporeal and the vegetal intentionalities, the subject/object dichotomy is irrelevant; their acts of living do not "objectivate" that toward which they orient themselves and therefore do not obey a strict ideal separation of *noesis* from *noema* in the expectation of a pre-delineated "fulfillment" of the intending in the intended. (Even assuming such fulfillment were plausible, it would have been fleeting and would not have exhausted the "empty" intentionality of growth in the presence of its nebulous *noema*, the light.) We are akin to plants in that, like them, we most often act without our heads, without irradiating commands from the central point of consciousness or the brain—and it is by no means evident that the brain itself is subject to this hierarchical centralization—all the while upholding a certain non-conscious logic and consistency in our acts of living. During a great portion of our lives, the vegetal *pas de tête* dictates the rhythm of human existence.

The intentionality of human pre-reflective acts is not automatic but rather existential, or, as Merleau-Ponty unambiguously states in a footnote to *Phenomenology of Perception*: "In our opinion Husserl's originality lies beyond the notion of intentionality; it is to be found in the elaboration of this notion and in the discovery, beneath the intentionality of representations, of a deeper intentionality, which others have called existence."[20] Does the existential character of human pre-reflective intentionality set it apart from that of plants? Not if we go a little further in the direction of phenomenological anti-humanism by contending that non-human existences also have their corresponding intentionalities, in some cases intersecting with or underlying our non-conscious existence. And so the intentionality of plants, similar to the pre-reflective comportment of the human, is seamlessly connected to its spatial, physical milieu, so much so that the abstraction of both from the environmental context, wherein they are embedded, risks irreparably disturbing and losing sight of them *qua* intentionalities.

While the intentionality of nourishment is easily demonstrable, in the case of reproduction the matter is more complicated and requires a further theoretical elaboration. Aristotle implicitly schematizes this part of vegetal intentionality in *De anima*, where reproduction is not an automatic "function," as the English translation of W. S. Hett makes us believe, but the "work" of vegetal soul, a vigorous and energetic *ergon* (415a26) setting its sights on multiple noematic targets. The reproductive intentionality of the plant is of course to "reproduce its kind" not for itself but for the species it belongs to. "That for the sake of which" such work is performed, the beneficiary of reproduction identified in proto-phenomenological terms prefiguring Husserl's philosophical project, is the genus that continually renews itself thanks to the production (*poiesis*) of new individuals.

But, Aristotle reminds us, "that for the sake of which" can also describe "that for the purpose of which," a deeper source of motivation and meaning approximating the final purpose "for the sake of which" everything is enacted, i.e., the Good that, in the last instance, inspires all living and thinking. The plant's reproduction does not culminate in that which is reproduced, be it a particular offspring—irrespective of how well it may fulfill the generative *telos* of the mother-plant, as Plato observes with regard to the first shoot that always sprouts with "excellence" (*Laws* 6.765e)—or an entire species; reproductive intentionality becomes interminable when it directs itself toward its ultimate target, *viz.* the immortal and the divine (*tou aei kai tou theiou*), in which it can participate in the only way it can, by giving rise to another like it (415a27–b9). The plant's destination, if it has one, is ethical; the Good is the ultimate form of its life.

So conceived, the "intended" instigates reproductive intentionality to carry on its work *ad infinitum*, because no instantiation of a particular plant or species is able to lay claim to the immortal and the divine as such. Our human soul also partakes of the immortal through reproduction, thereby sharing in the intentional activity of plants, though this is not the only possible route we might take toward immortality and divinity (for Aristotle, *theoreia*, or "thought thinking itself," is of course a surer path, leading toward the same destination). That the reproductive intentionality of plants is the material precursor to the purely theoretical acts of thinking becomes evident already in the Socratic analogy between bodily generation and the birthing of ideas, grounded, in more or less sublimated ways, in the generative function of the vegetal soul. In light of this common root, the material reproduction of the body turns into a prototype of thought, while the plant's intentionality comes to denote the most concrete mode of thinking imaginable. Pursuing this line of reasoning, the consciousness-centered intentionality that preoccupies traditional phenomenologists will find a broader application if it surpasses the narrow parameters of anthropocentrism and embraces a multiplicity of non-conscious existences, including that of plants.

Thinking Without Identity

Vegetal being revolves around non-identity, understood both as the plant's inseparability from the environment wherein it germinates and grows, and as its style of living devoid of a clearly delineated autonomous self. Our ongoing efforts to render the categories of vegetal being in the terms of plant-thinking cannot disregard this important facet of ontophytology.

The most obvious symptom of the plant's non-identity is its unrest, reflecting the plasticity and restlessness of life itself: its ceaseless striving toward the other and becoming-other in growth and reproduction, as well as in the metamorphosis of these vegetal qualities into human and animal potentialities. To attribute static identity to the plants' way of being and thinking is therefore to disregard their very vivacity. But this is exactly what seems to be going on in the correlation Nietzsche draws between the plants' repose, which he assumes to be exhaustive of their mode of being, and an identitarian thinking inspired by them and said to presage the formal logical approach to the world.

In a fragment from *Human, All Too Human*, ominously titled "Fundamental Concepts of Metaphysics," Nietzsche writes: "To the plants all things are usually in repose, eternal, every thing identical with itself. It is from the period of the lower organisms that man has inherited the belief that there are *identical things*. . . . It may even be that the original belief of everything organic was from the very beginning that all the rest of the world is one and unmoving."[21] This assertion, most likely meant to scandalize logicians by conjecturing that they are the direct legatees of beliefs prevalent in "the lower organisms," hinges on a double repression of vegetal temporality: besides imputing to plants an incapability to experience the passage of time, Nietzsche proves to be impervious to their constant alterability, which Goethe and Hegel emphasized before him. On Nietzsche's view, then, humans in the state of repose and non-sensation are temporarily indifferent to the world and "notice no alteration in it," but the plants are permanently unperturbed, existing as if their environment were unaltered and "eternal."[22]

In this regard, an empirical question we could pose to Nietzsche is whether the plant still notices no alteration in the world when its biosphere is drastically changed, for instance as a result of drought, toxic pollution of the soil in which it is rooted, a plague of insects, or other catalysts. The creative mutual interaction of any living being and its environment, on the one hand, rules out such absolute insensitivity and, on the other, substantiates vegetal thinking devoid of identity and encompassing the plant along with its biosphere. If the logical belief in identical or self-identical things really stems from the prehistory of the human, then one must search for its source in what preceded vegetation, that is, in the inflexible, inorganic world of minerals, where, too, this belief would not be entirely justified.

We may excuse or, at the very least, explain Nietzsche's theoretical violence against plants with recourse to the Pongean definition of vegetal beings as "living crystals," hinting at their ontological proximity to and, at the same time, decisive modification of the inorganic realm. This qualified approximation, on the plane of plant-being, to the world of minerals cannot help but have a significant impact on the epistemic milieu of vegetal life. To appreciate the complexity of the Nietzschean "biology of the drive to knowledge," then, a brief fragment from *The Will to Power* must supplement the one I have extracted from *Human, All Too Human*. In 1885 Nietzsche writes in a shorthand: "'Thinking' in primitive conditions (pre-organic) is the crystallization of forms, as in the case of crystal.—In *our* thought, the essential feature is fitting new material into old schemas (=Procrustes' bed), *making* equal what is new."[23] The stability and identity previously pinned on plant-thinking are here unequivocally relegated to the pre-organic "crystallization of forms" that survive in human thought in the shape of Kantian immutable categories and forms of intuition to which all novel experiences must in one way or another conform.

If the plant is a "living crystal," then in its being, as well as in its thinking, it enlivens this pre-organic heritage, putting in motion—which is to say, *de*-formalizing, undoing, or de-constituting—the inflexible "old schemas." The event of what is new, what is irreducible either to previous experiences or to the empty transcendental molds for processing them, is first intimated in plant-thinking, which destroys the Procrustean bed of formal logic and transcendental *a priori* structures—those ideal standards to which no living being can measure up fully. Although it hovers between pre-organic proto-thinking and "*our* thought" (which has imbibed the anachronistic methods, if not the conclusions, of the latter), plant-thinking supersedes subsequent cognitive-evolutionary developments, to the extent that, instead of "*making* equal what is new" and what is old, it facilitates the coming to pass of the event, of that which is unforeseeable, because irreducible to the schemas of the past. It stands for a thinking that admits difference into its midst and operates by means of this very difference, consonant with the ontology of plants.

The non-anthropocentric thinking of difference no longer fitting into the schemas of identitarian thought may not be recognized for what it is; it may lose the familiar outlines of epistemic systems as they have been theorized in the history of philosophy. This non- or misrecognition is not an accident. Mirroring the plants' heteronomy, its ontological dependence on something other than itself, such as the light, plant-thinking is so closely entwined with its other (i.e., with non-thinking) that it does not maintain an identity *as thinking*. It rejects the principle of non-contradiction in its content and in its form, in that, at once thinking and not thinking, it is not at all opposed to its "other."

Aristotle chanced upon this same insight in *Metaphysics* (1008b10–11)—"If one has no belief of anything, but is equally [*homoiōs*] thinking and not thinking, how would one differ from a plant?"—where he reiterated the insulting comparison of someone who does not respect the tenets of formal logic made earlier in the text (1006a12–15). In Aristotle's view, a human being equal (*homoiōs*) to a plant is one who is equally (*homoiōs*) thinking and not thinking; the erasure of the difference between "A" and "not-A," which is a *de facto* violation of the principle of non-contradiction, cancels out the onto-metaphysical difference between the human and the plant. Epistemological reality defines ontological existence, so that manners of thinking determine modes of being well before the advent of German idealism. When certain ways of thinking happen to be inappropriate to the being that employs them, they interfere with the ontological makeup of this very being, introducing a correction and making who or what we are fit the way we think. The human who thinks like a plant literally becomes a plant, since the destruction of classical *logos* annihilates the thing that distinguishes us from other living beings. In response to Deleuze and Guattari's injunction, "Follow the plants!"[24] we will engage in irreverent plant-thinking, which will set us on the path of becoming-plant.[25]

To be fair, a vegetable-like person is not one who no longer thinks but, in a more nuanced formulation, one who thinks without following the prescriptions of formal logic and therefore, in some sense, without thinking. Let us then try to get accustomed to the idea that thinking is not the sole prerogative of the subject, or of the human being, and that, aside from altering the form of thought (which becomes inseparable from its opposite, the non-thought) and changing its content (which includes contradictions), "non-identical thinking" indicates freedom from the substantive and self-enclosed identity of the thinkers themselves. In place of the Kantian transcendental synthesis of *I think* that supposedly accompanies all my representations, plant-thinking posits *it thinks*, a much more impersonal, non-subjective, and non-anthropomorphic agency. But who or what is the "it" that thinks?

The "it" that thinks is the plant, a being whose self, in Hegel's formulation, is to be found in the other, to which it strives. Therefore, whenever *it thinks*, what does the thinking is it itself and, at the same time, its other, what it is *not*. How is this possible? In what follows, I would like to glean three modalities of the vegetal "it thinks" from twentieth-century philosophers Bergson, Gregory Bateson, and Deleuze.

Bergson's *Creative Evolution* as a whole broadens the sphere of the intellect, redirecting it from the self-identical "facts" it seeks toward life processes, and simultaneously restricts this sphere to one among many instances of active evolutionary inventiveness. As in the rest of his philosophy, Bergson encourages the kind of thinking that thinks *with* life, not *against* it. Whether it has to do with the plant or the human, *it thinks* points toward the thinking of life itself, a de-formalizing activity that, when inserted into the categories of conceptual thought, implodes them from within: "In vain we force the living into this or that one of our molds. All the molds crack. They are too narrow, above all too rigid, for what we try to put into them."[26] Stretched on the Procrustean bed of logic, the living cannot be made equal to the form and content of our cognitive molds; the thinking of life is in and of itself a thinking of non-identity, unsettling the human intellect, which, left to its own devices, "feels at home among inanimate objects, more especially among solids," so that "our concepts have been formed on the model of solids" and "our logic is, pre-eminently, the logic of solids."[27] Bergson finds himself in a tacit agreement with Nietzsche: the intellect's crystalline, crystallized structure, having congealed at the pinnacle of modern philosophy into the Kantian *I think*, is dead thought, but this thought will be de-solidified, enlivened, and transformed into *it thinks* as soon as it endeavors to "digest" the life processes that do not rest in a final identity with themselves.

The life that thinks, be it through us or through the plant, is a far cry from an undifferentiated flux of becoming, a vortex of immanence sweeping everything into its homogeneous mix. The living-thinking of life is appropriate, in each case, to the relation of a given organism to its milieu. The role of our intellect, enunciated in this way, is to "secure the perfect fitting of our body to its environment,"[28] not by indulging in egoistic adaptation at any cost but by creating a unified ensemble of this body and its world. The philosophical sense of the Bergsonian "fitting" is unmistakable, for rather than repeating the traditional evolutionary mantra of the "survival of the fittest," it harkens back to the ancient Greek notion of the "fit" as a matter of appropriateness, adjustment, and ultimately justice.

What befits the life of a plant in its environment and what shapes plant-thinking, exercised by the plant and its other (that is to say, its milieu) as a single unit, is not the same thing that is appropriate to the integrated thinking of the human being and *its* life-world, though, due to the role of plant-soul in making a shared life possible, one may expect certain overlaps between the two kinds of intellection. It is the exigency of life in the midst of organic nature that such a fit be continually reconfigured, fine-tuned, and readjusted, because immutable and solidified concepts are useful only for orienting us in an environment made entirely of steel and blocks of concrete. Plant-thinking performs this function for the plant, suiting it to its milieu, which is to say, to itself *qua* other. The issue of environmental justice, conceived in the ancient sense of *dikē* (which, as Heidegger reveals in his reading of Anaximander, names in the same breath a jointure or a juncture), thus delineates the horizons of plant-thinking, conjoining the plant and its other.[29]

In a programmatic text, titled "Steps to an Ecology of Mind," Bateson underscores the epistemic consequences of this jointure, which, if thought through to its logical conclusion, implies that the "unit of survival is *organism* plus *environment*."[30] The "it" that thinks is both more and less than the "I." More, because it is incapable of thinking by means of a mere "I" divorced from the environmental component of the unit of survival. Less, because this unit is neither as individuated nor as autonomously separate as the subject of thought. Whereas the plant is fully embedded within the holistic mode of thinking and being invoked by Bateson, the human sets itself over and against its environment, driving a wedge in the unit of survival, wherein it participates. The ensuing disjointure or disadjustment heralds, in addition to calamitous environmental injustice (*adikia*), the impossibility of the organism's continued existence; in the very moment of asserting and celebrating its unique power and autonomy, it undermines itself in virtue of persecuting and destroying the other within and outside of itself. This is what in modern philosophical parlance is called "alienation": an ontological condition replete with detrimental epistemological effects, including insanity. If the environment, along with which you form a unit of survival, is Lake Erie and if "you decide that you want to get rid of the by-products of human life and that Lake Erie will be a good place to put them," "Lake Erie is driven insane [and] its insanity is incorporated in the larger system of *your* thought and experience."[31]

In the face of the insanity of transcendent thought, plant-thinking, immanent to the milieu wherein it thrives, will be the signpost of, or a concrete normative ideal for, the Batesonian version of *it thinks*. It will permit us, among other things, to read with fresh eyes the famous quip of Pascal, "Man is a reed, the weakest in nature, but he is a thinking reed."[32] The thinking of this reed is precisely what makes it weak, emasculates its integral connection to the environment, prompts it to harm itself and the surrounding world. We might, however, imagine a different kind of weakness that would be associated with the thinking of the human reed and would come about as a result of realizing its frailty, the fragility of its milieu and of the conjunction (the "plus") at the heart of Bateson's "unit of survival." This realization takes us a step closer to post-metaphysical thought. Mitigating the excessive separation of the human mind from the context of its embeddedness, non-oppositional plant-thinking will therefore be entrusted with guarding the sanity of our thought and with maintaining it adjusted to our life-world. A guarantor of environmental justice, the vegetal *it thinks* will moderate the lethal tendencies of the human *I think*, neglectful of the non-individuated foundations of thought and of the context integral to its formalization. In a paraphrase of Heidegger, it is not a god but a plant that can save us.

Deleuze and Guattari, who have relied extensively on the philosophies of both Bergson and Bateson, similarly privilege vegetal heteronomy and hetero-affection in plant-thinking. They write: "The wisdom of plants: even when they have roots, there is always an outside where they form a rhizome with something else—with the wind, an animal, human beings (and there is also an aspect under which animals themselves form rhizomes, as do people, etc.)."[33]

The third instantiation of the *it thinks* is the rhizome, which, instead of opposing, supplements its other and which accomplishes the work of the vegetal soul, traversing metaphysical distinctions between plants, animals, and human beings. Rhizomatic thinking is the thinking of exteriority in and as exteriority, the inextricable relation to "an outside," to something other, including parts of inorganic nature, other living beings, and the products of human activity. Its non-identity, in the writings of Deleuze and Guattari, reproduces the relational character of

Bateson's eco-mental systems and of Bergson's "fitting" of the body to its environment, so that the organism and elements of the biosphere to which it belongs form nodes within the forever-unfinished mesh of the rhizome. Rhizomatic thought—or plant-thinking *proper*—takes place in the interconnections between the nodes, in the "lines of flight" across which differences are communicated and shared, the lines leading these nodal points out of themselves, beyond the fictitious enclosure of a reified and self-sufficient identity. The vegetal *it thinks* does not answer the question, "Who or what does the thinking?" but, "When and where does thinking happen?" because this thinking, inseparable from the place of its germination, arises from and returns to the plant's embeddedness in its environment. All radically contextual thought is a worthy inheritor of vegetal life, which continues to thrive, proliferating, among other places, in those texts that lay bare and reveal their own margins; hermeneutics, historicism, immanent criticism, and deconstruction are the methodological names for this inheritance.

A preliminary response to the question of the lived spatio-temporal conditions of thought is that plant-thinking happens (1) *when* the presumed self-identity of "subjects" and "objects" that populate a given milieu recedes, allowing the rhizomatic assemblage to surge up to the foreground, to be activated by sharing difference among its various nodes, and (2) *where* the spacings and connections, communication lines and gaps between the participants in this assemblage prevail over what is delimited within them. If this image of thought is evocative of the synapses, whose firing accounts for the brain's neural activity, then we must conclude that the brain is a neurological elaboration on the decentered vegetal *it thinks*: "The discontinuity between cells, the role of the axons, the functioning of the synapses, the existence of synaptic microfissures, the leap each message makes across such fissures, make the brain a multiplicity immersed in its plane of consistence or neuroglia. . . . Many people have a tree growing in their heads, but the brain itself is much more a grass than a tree."[34] When *it thinks*, it does so nonhierarchically and, like the growing grass, keeps close to the ground, to existence, to the immanence of what is "here below." The competing vegetal modulations of the brain, transposing either a top–down tree structure or a horizontal grass layout onto neural organization, are in any event beholden to plant-thinking, which induces the non-identity of human thought, prompted to mold itself in the likeness of what it is not, namely the plant and *its* thinking. At the core of the subject, who proclaims: "I think," lies the subjectless vegetal *it thinks*, at once shoring up and destabilizing the thinking of this "I."

Philosophy, a Sublimated Plant-Thinking

Between the eighteenth and the twentieth centuries, the Aristotelian capacities of the vegetal soul—to obtain nourishment and to reproduce—received a new lease on life. The significance of this revival can be hardly overestimated, since it has culminated in a discovery of the direct involvement of vegetal intentionality in sensation and cogitation, i.e., those parts of the psyche that, according to *De anima*, pertain to the souls of animals and humans, respectively.

Earlier I commented on the importance of digestion for Nietzsche, who is a veritable physiologist of thought, acutely aware of the way the "lower" functions of the body bear upon the highest expressions of spirit. Not only are diets, linked to the nutritive function of plant-soul, responsible for the style and content of our thought, but also the climate in which we live is determinative for the development of culture, or the sum total of Spirit (tropical climate, for instance, gives rise to "violent antitheses, the abrupt transition of day to night and night to day, heat and vivid color, reverence for everything sudden, mysterious, terrible", etc.).[35] Setting aside the issue of whether Nietzsche is sufficiently careful in his deduction of causal relations binding food and climate, on the one hand, and cognitive and cultural orientations, on the other, the effort at re-embedding thought and culture in their material conditions is a nod of acknowledgment to vegetal life, heteronomously regulated by elements in its own milieu.

Despite the compelling nature of Nietzsche's contribution, it is Novalis who is, perhaps, the most explicit plant-thinker in modern philosophy. In his exposition of sense, Novalis purposefully deploys vegetal imagery and language—"Sense in general eats, digests or fecunds, conceives—is fecundated by light"[36]—at the point of convergence of the nourishing and reproductive capacities of plant-soul ("digests or fecunds"). It is as though, in sensation, these capacities are elevated to a higher spiritual sphere, sublimated, and idealized, notwithstanding their being tethered to the vegetal source. The most ideal and luminous of the senses—vision—finally gets in touch with that to which the plants tend as well: it "fecunds" and "is fecundated" by light, without which it could not fulfill its function. Despite the celebrated ideality of vision, it, like all the other senses, is engrossed in the materiality of digestion, in nutritive activity that does not spare materiality as a whole, digested into the world of Spirit. Sublimation is a matter of digestion, an overarching vegetal *dunamis* that regulates, among other things, the transformation of the plant-soul into its sensory and cognitive counterparts.

Nor is sensuousness, or enjoyment, spared the logic of digestive assimilation, given that "all enjoyment, all taking in and assimilation, is eating, or rather: eating is nothing other than assimilation. All spiritual pleasure can be expressed through eating. In friendship, one really eats of the friend, or feeds on him."[37] There is but one crucial difference between vegetal assimilation and its spiritual permutation: in the absence of interiority, the former assimilates the plant to its other, whereas the latter appropriates the other to itself. Imagine then a way of thinking where thoughts or discernments are not stored in the interiority of consciousness—or else, a sublimated stomach—but circulate on the surface and keep close to the phenomenal appearances of things. This image of thought will not sound bizarre to those familiar with the basic insights of phenomenology, which, in addition to denying the existence of noumenal reality behind the curtain of appearances, lambastes the view of consciousness as an interior drawer for the storage of thoughts and for the memories of past experiences. Jointly, the privileging of light in its account of knowledge and the essential superficiality of phenomena put phenomenology on the side of plant-thinking—the epistemophytological articulation of vegetal ontology.

Hegel concedes that the act of devouring a thing is "the most elementary school of wisdom [*Schule der Weisheit*]," from which the animals are not excluded and which, we should add, is predicated upon the nutritive capacity of plants.[38] But it is by no means certain that the German philosopher himself has ever graduated from what he disparagingly calls "elementary school." At every stage of the dialectic, the assimilation of the object, devoured by Spirit, signals the resolution of a particular standoff and instigates a gradual progression of Spirit from implicit consciousness to absolute knowing. Of course, dialectic-generating resistance may emanate from an external object, or it may derive from the I as an object related to itself in self-consciousness. But, whatever its precipitating factor, each transition to a higher stage is inconceivable without a more effective assimilation, consumption, and consummation of the obstacle in the actualized interiority of Spirit. There are no qualitative discontinuities between acts of eating and thinking, between successfully accomplished mediations of the subjects of need, desire, understanding, self-consciousness, and so on with their corresponding objects, because all these acts belong under the common spiritual aegis of assimilation. Everything Real becomes Rational, as a result of the Rational swallowing up, digesting, and manifesting again, in a regurgitated form, the previously unmediated Real. Were it not for the dialectical hypostatization of the principle of interiority, foreign to vegetal life, we could say that the repeated sublimation of nutritive capacity in Hegel's texts is tantamount to a myriad of ways of elaborating upon plant-being and plant-thinking.

On the itinerary toward absolute knowing, whereby Spirit will have recognized itself as Spirit, plant-thinking both orchestrates and delimits the process of assimilation; although various parts of plants are easily turned into food, the vegetal principle of nourishment, presiding, in a disguised form, over the dialectical process as a whole, is indigestible and inassimilable. Deconstructive reminders concerning that which cannot be consumed, digested, or indeed deconstructed—reminders that, in the last instance, put the subject face to face with the question of justice—are signs of respect to the absolute material resistance inherent in vegetal life. Barely recognizable, sublimated and sublime, *to threptikon* regulates all nutritive processes, so that to consume or to digest it would be, still, to follow its precepts.

When in a 1990 interview Daniel Birnbaum and Anders Olsson raised the question of the parallels between deconstructive reading and a certain style or a manner of eating, Derrida responded, "[A deconstructive reading] would mean respect for that which cannot be eaten—respect for that in a text which cannot be assimilated. My thoughts on the limits of eating follow in their entirety the same schema as my theories on the indeterminate or untranslatable in a text. There is always a remainder that cannot be read, that must always remain alien."[39] This remainder is what, approximately twenty years earlier, in *Glas*, Derrida had designated as "morsels," those obstinate leftovers that could not find their proper place within the scope of Hegel's system and that stand for material obstacles to the routines of idealization, rational comprehension, and conceptualization.[40] Faithful to the obscurity of vegetal life, plant-thinking preserves the unthinkable in its midst. It insists, in Hegelian terms, on the imperviousness of a sizeable portion of "unconscious Spirit" to Spirit conscious of itself. Like the plant, it is only partly exposed to light, since its roots are immersed in the moist darkness of the earth, in non-comprehensible materiality, and in subject-less, object-less intimacy tending toward the abolition of distance.

The double impossibility of describing mental processes in an objective fashion and becoming a fully conscious subject explains, from a philosophical point of view, the interminable nature of psychoanalytic interpretation. Those who pay heed to Freud's theory concerning the origins of human knowledge in the "early sexual researches" of a child will discern in it the other half of plant-thinking, now clustered around the reproductive capacity of the vegetal soul as it survives, in an altered state, in human beings.[41] The take-home message of *Three Essays on Sexuality* is that all knowledge—not to mention intellectual curiosity, the drive or the desire to know—arises from our embodied ontology, psychoanalytically codified as "sexuality." Nowhere is this more clearly or forcefully expressed than in the section of the *Essays* titled "The Sexual Researches of Childhood." Let us, then, consider the unacknowledged vegetal background of human knowledge on the basis of this text.

The "epistemophilic drive" is Freud's name for the attitude of intellectual curiosity indicative of the love of knowing, which is to say, philosophy in its simplest and most sensuous form. The desire to know first awakens in response to two basic mysteries confronting the child: the question of sexual difference and the riddle of where babies come from. These inquiries stand at the threshold of all epistemic activity "since we have learnt from psycho-analysis that the instinct for knowledge in children is attracted unexpectedly early to sexual problems and is in fact *possibly first aroused by them*."[42] Much in the subsequent thinking of the young sexual-researcher will depend on the outcome of these initial investigations, responding to the provocation of the vegetal *dunamis* in us.

Although the search for answers is the child's first timid flirtation with independent thinking, for the most part it fails, culminating in the reduction of sexual difference to underlying sameness. How does one grapple with this failure? As a result of the disavowal of sexual difference, castration anxiety germinates in the hypothesis that women are but deficient men and that therefore the distinction between the sexes overlays their prior substantial homogeneity. Such dismissal of sexual difference provokes the leveling and negation of all difference. "It therefore follows," Freud writes, "that the efforts of the childish investigator are habitually fruitless, and end in a renunciation which not infrequently leaves behind a permanent injury to the instinct for knowledge."[43] The epistemophilic drive suffers irreversible damage, depending on how frustrating the early researches have turned out to be. What determines the fate of independent ideation is this early contemplation of human sexuality and especially the crude theorizing about the mystery of reproduction, the principle of vegetal existence shared by all living beings. Thinking as such commences, necessarily, as plant-thinking and it may come to an abrupt end when the obscure vegetal grounding of the "life of the mind" is withdrawn, its attendant differences disavowed.

Given the deep affinity between our cognitive apparatus and the reproductive function of the vegetal soul, we may graft the psychoanalytic structure consciousness/unconscious (*Cs./Uncs.)* onto divisions within the Aristotelian psyche, furnishing yet another piece of evidence for the provenance of human thought from plant-thinking. On the flip side of this recognition, we diagnose a powerful tendency to repress the questioning of our relation to plants, alongside the question of sexual difference, in statements disowning the unconscious roots of consciousness. Historically, the occidental-metaphysical denigration of vegetal life has been a symptom of repression, an acting-out of the fact that humanity has not yet come to terms with its other-than-human heritage.[44] The repression of sexuality inflicts a serious wound upon knowledge: it triggers much more than either a simple devaluation of corporeity within the infamous "mind–body split" or a forgetting of human finitude. The tragedy is that it also distorts our relation to the environment (or better yet, to environments, the worlds of other living beings, such as plants) and, in so doing, prejudices our very capacity for survival.

In retrospect, Freudian findings may explain an otherwise cryptic fragment by Novalis that "thinking, like flowering, is but the most delicate evolution of plastic forces—the universal power of Nature elevated to the potency of the *nth* degree."[45] Flowering is a sign for the sexual maturation of the plant, just as thinking is the token of human ripeness (with the psychoanalytic proviso that it—thinking—is nourished, in the first instance, by embodied reflections on sexuality); the "universal power of Nature" they have in common is the vegetal capacity for reproduction, which, in virtue of its plasticity, is sublimated into abstract thought. Repression interferes with and blocks this power's transformation into a still higher potency, arresting this "delicate evolution" and militating against the very possibility of possibility, that is, the dissemination of pollen or sense.

The primacy of the sexual function in psychoanalysis implies a clear-cut choice between the two faculties of the vegetal soul. It means that Freud divides the psyche along the classical lines already drawn in antiquity; that he accords more significance to the reproductive capacity than to the nutritive *dunamis*; and that, at bottom, he imbues acts of eating with sexual meaning. Not only is conceptual assimilation equivalent to a refined form of eating, but also the act of devouring food is saturated with oral sexual pleasure: "The satisfaction of the erotogenic zone is associated, in the first instance, with the satisfaction of the need for nourishment."[46] Psychoanalysis exposes the hidden foundations of the Hegelian "elementary school of wisdom," when it comes up with the thesis that there is no eating up the other without a derivation of sexual pleasure in the process, and that furthermore there is no knowing without this sublimated orality. Whether it is the logic of nourishment or the reproductive function that is considered primary, something of the vegetal soul in us accounts for the flourishing of thinking. Human cogitation is born of the non-conscious intentionality of plant life, which accords with the psychoanalytic axiom that much of our psyche is swathed in the unconscious, the medium of vegetal existence.

In Western philosophy, the transition from the ignorance of the unconscious to conscious existence has been portrayed as an emergence from darkness into the light of knowledge.[47] In the seedling's sprouting from the soil and striving to the light of the sun, philosophers as diverse as Plato and Hegel saw the natural precursor to human education, while eighteenth- and nineteenth-century German thinkers detected what Meyer H. Abrams would later term the "vegetable genius."[48] Nietzsche brands metaphysical philosophers "rare plants"—and it is desirable, he notes, that they keep it that way[49]—not only because of the empirical paucity of their numbers throughout history but also because, in contrast to all other plants, they seem to evade that radiance which emanates both from the literal sunlight and, in the case of human beings, from the alluring fluorescence of myth. Both vision and mythical thinking are points of access to an illusory reality, from which the philosopher wishes to flee: "Now, the Greek philosophers deprived themselves of precisely this myth: is it not as if they wanted to move out of the sunshine into shadow and gloom?" "But," Nietzsche continues, "no plant avoids the light; fundamentally these philosophers were only seeking a *brighter* sun, the myth was not pure, not lucid enough for them."[50]

The light of Ideas, toward which the philosophical soul grows as though it were an etherial plant striving toward the sun, supersedes in its clarity and brilliance physical light, with the principle of "spiritual" growth not diverging from but rather modifying vegetal proliferation. In Platonism, the "*brighter* sun" is also all the warmer, in that its eidetic luminosity is still related to the kind of warmth that is generative and creative, allowing beings to spring into being. Needless to say, this life-giving heat of philosophical heliocentrism is absent from the thinking of the Enlightenment, which analogizes reason to neutral light, capable of coldly and dispassionately illuminating everything, and from twentieth-century phenomenology, preoccupied with the infinitely varying modes of appearing, with how things come to be, come into the light, are illuminated with meaning. Nevertheless, what unites the three milestones of Western thought is the way they put vegetal movement toward light in the service of our thinking about thinking, a meta-theorizing about human knowledge.

The relation of plant-thinking to Platonism, the Enlightenment, and classical phenomenology is ambiguous, to say the least. Above all we must be cognizant of the possibility that the search for a brighter sun would threaten, at any moment and right in the midst of this marvelous luminosity, to devolve into the new "Dark Ages," where the fully conscious and self-conscious existence brutally represses the unconscious remainder it cannot do away with; where vegetal life—and, along with it, everything belonging in the sphere of immanence—undergoes a thorough enucleation both within and outside the human subject; and where such repression of darkness severs the intellect, seeking pure light, from its roots swathed in obscurity. The limit to the kinship of traditional thought with vegetal proliferation is precisely this: the metaphysical project bent on leaving the darkness of mere life behind undercuts the conditions of its own existence (or of any existence, for that matter). Incapable of acknowledging the thinking coextensive with the variegated acts of living, metaphysics wields the power of negativity and death even when it seems to be growing toward another kind of light and to affirm the quasi-divine life of the mind. An excrescence of plant-thinking, it nonetheless risks turning into a cancerous growth, suffocating the very entity from which it draws its vitality.

If plant-thinking is to avoid being caught in the trap of the preceding metaphysical strategies that selectively inherited and at the same time violated vegetal life, it must be receptive to and appreciative of this life's other pole, the pole of darkness with the possibilities proper to it. In the words of Lev Shestov, "It seems that, very soon, human beings will feel that the same little-understood but caring force, which has thrown us into this world and taught us, like the plants, to tend toward the light, gradually readying us for a free life, is prodding us toward a new sphere, where a new life with its own riches awaits us. And, perhaps, in the not-so-distant future, an inspired poet . . . will courageously and joyfully exclaim: 'Let the sun disappear, and let there be darkness!'"[51] Plant-thinking is obliged to undersign the desire of Shestov's "inspired poet," to the extent that it reconnects with its unconscious roots, all the while refraining from the indiscriminate repudiation of light. To live and to think in and from the middle, like a plant partaking of light and of darkness, is not to be confined to the dialectical twilight, where philosophy paints "its grey on grey." It is, rather, to refashion oneself—one's thought and one's existence—into a bridge between divergent elements: to become a place where the sky communes with the earth and light encounters but does not dispel darkness.

Epilogue

The Ethical Offshoots of Plant-Thinking

E as palavras? Aonde vão? Quantas permanecem? Por quanto tempo? E, finalmente, para quê? . . . Penso que meu avô Jerónimo, nas suas últimas horas, se foi despedir das árvores que havia plantado, abraçando-as e chorando porque sabia que não voltaria a vê-las. A lição é boa. Abraço-me pois às palavras que escrevi, desejo-lhes longa vida e recomeço a escrita no ponto em que tinha parado. Não há outra resposta.
—José Saramago, "86 anos," *Caderno azul*[1]

Look at the flowers, keeping good faith with the world,
to whom we at destiny's margins presume to lend destiny.
Who can be sure they do not regret how they fade?—
for to be their regret is perhaps our own duty.
—Rainer Maria Rilke, "Sonnet XIV," *Sonnets to Orpheus*, part II

In 2008 the Swiss Federal Ethics Committee on Nonhuman Biotechnology released a report titled "The Dignity of Living Beings with Regard to Plants." In this document, perhaps for the first time in human history, a government-appointed body issued recommendations for the ethical treatment of plants, or, as the subtitle of the report indicates, for the "moral consideration of plants for their own sake." The Swiss Committee reached an unprecedented agreement: that vegetal life not only deserves to be treated with the kind of dignity extended to all other living beings but that it also possesses an absolute moral value, irreducible to the instrumental rationale behind efforts to protect biodiversity and to enhance "conservation." Henceforth, subjecting plants to "arbitrary harm" would be considered morally reprehensible, while the actual instances of instrumentalizing these living beings would require moral justifications.

Although the revolutionary potential of the report was undeniable, it failed, at its foundations, to inquire into the being of plants, into their unique purchase on life and thus into what sustains their dignity under the heading of constitutional law protecting *die Würde der Kreatur*, translated as "the dignity of living beings." In other words, what is missing from this otherwise admirable document is an ontological point of departure questioning the very category of "plants" and the problematic term "living beings." This oversight carries a host of epistemological and ethical consequences. On the epistemological side of things, supplementing the silently presupposed ontological status of vegetal life, we find a strong emphasis on the rational scrutiny of moral intuitions, buttressed with technical "decision trees" meant to guide all subsequent discussions concerning the status of plants. But even when the collective knowing subject, comprising the report authors, ostensibly refuses to relinquish its own epistemo-metaphysical privilege as a moral arbiter, rational analysis and the decision trees it relies upon undermine themselves, come internally undone, insofar that they culminate in a "morally relevant" profession of not-knowing, in response to the question of the intrinsic Good as it pertains to plants.[2] Ethically, too, the designation of plants as "moral objects,"[3] molding them to Kantian and post-Kantian philosophical discourses, already puts them at the disposal of the human subject well in advance of any decision on their status. Still, the committee's debate on the meaning of the vegetal moral object, which could be an individual plant or plant collectives—i.e., reproductive plant communities—betrays its members' unease as to the ontological correlate to their freshly minted moral injunction.

To skirt the pitfalls of efforts aimed at bringing the entire vegetal world into the fold of moral philosophy (and, surreptitiously, of metaphysics), it is *de rigueur* to cultivate a way of thinking not only *about* plants, understood as epistemic or moral objects, but also *with* them and, consequently, *with* and *in* the environment, from which they are not really separate. What is required therefore is the cultivation of a certain intimacy with plants, which does not border on empathy or on the attribution of the same fundamental substratum to their life and to ours; rather, like all intimacy, it will take place (largely) in the dark, respectful of the obscurity of vegetal life.

We are already acquainted with the contours of the enterprise that encourages us to become plant-like in our thought: avoiding the subject/object split, which dogmatically separates us from what we aspire to know; pursuing the trajectory of intentionality's atelic dispersion; becoming a passage or a medium for the other, and so forth. But exactly how does plant-thinking reflect and bear upon the ethical treatment of plants? In what follows, I will describe ten offshoots of this thinking, each of them addressing the above question.[4]

The first offshoot: Plant-thinking is *plant-doing*. Vegetal life practically deconstructs the metaphysical split between the soul and the body, eliminating, in the same gesture, the classical opposition between theory and practice. The material, extended, and organic character of vegetal thought, grounded in a particular mode of being shared by all living creatures, is at one with the activities of nourishment and generation. In this sense, plant-thinking is active through and through; it has no practical effects because in and of itself it is a *habitus* of living. All the subsequent offshoots will be grafted onto this insight and its core implication: that our engagement in plant-thinking actively takes the side of the plant and works for the sake of the plant. No neutrality, no objectivity—only the proliferation of vegetal life, in plants as well as outside of them, through the mutually supplementary dimensions of extended thinking and doing.

The second offshoot: Ethics as such is an offshoot of plant-thinking. If ethics, understood *à la* Levinas, is the relation to the other, then it must be rooted in the ontology of vegetal life, heteronomously defined by a striving to alterity. Environmental ethics or, more precisely, the ethics of plants is therefore a kind of homecoming, a harkening of ethical discourse back to the domain of life wherein it originated. Consequently, vegetation cannot be treated merely as an object, moral or otherwise, since it is also an agent, if non-active, related to its other (inorganic nature, light, etc.)

The third offshoot: Vegetal life deserves respect. It is not sufficient to conclude, as the authors of the Swiss report do, that plants have an "intrinsic worth," especially because, at the dusk of metaphysics, the very division between the intrinsic and the extrinsic, the "in-itself" and the "for-us," no longer applies. Broadly speaking, what is "worthy" in vegetal life is that it embodies the an-archic principle of living and thereby constitutes plant-thinking (or, for that matter, plant-doing) with its acts of nourishment and generation applicable to all other living beings. Vegetal life enlivens plants, as well as, in different ways, animals and human beings; the common life at its barest, it is in equal measure an end-in-itself and a source of vitality for-us. An offense against vegetal life harms both the plants we destroy and *something of the vegetal being in us*. Besides annihilating the plants themselves, the highly aggressive extermination of the flora, which has currently put under the threat of extinction up to one-fifth of all plant species on the planet, impoverishes a vital element in what we call "the human." Likewise, within the flexible notion of thinking I have outlined above, the more specialized Kantian idea of respect is appropriate to vegetal life. For Kant, human persons and other actually or potentially existing rational beings must be respected. Plant-thinking, of course, boasts non-rational reason of its own (the reason without which human rationality would not have been possible), even as elements of vegetal life ingress into the existential domain, usually reserved to human beings alone, and encompass temporality, freedom, and wisdom. Thinking and not-thinking, personhood and thinghood, are no longer pitted against each other in an oppositional relation but tightly interlaced in the plant, which has absorbed everything Kant deems worthy of respect, not to be confused with a quasi-religious veneration. How respect is to be paid to vegetal life is the question that we will touch upon in the other ethical offshoots germinating in these pages.[5]

The fourth offshoot: The plant is at once the most singular and the most general being; ethical concerns with vegetal life therefore pertain to plant ontology in each of its expressions and as a whole. In light of the plants' non-individuated and non-organismic existence, it behooves us to treat them as singularities, not as examples of a particular genus or species. Vegetal singularities, like the Spinozan *potentia*, are at the same time sub-individual and super-individual, consisting of plant "parts" devoid of an organic whole and of plant-communities (a single plant is a community, too) that are not equivalent to a collection of individuated members. No ethical approach to plants can evade this ontological paradox, itself a consequence of the plural and disseminated vegetal being. Whenever a plant or one of its parts is at issue, the plant-community and its entire milieu are affected, and vice versa. The "object"—if this term were still appropriate—of vegetal ethics is not one; it is both less than one and more than one at the confluence of the most singular and the most general. Such splitting of the ethical regard is indeed unavoidable. Our concern with a particular plant cannot afford to take place at the expense of the entire ecological community wherein vegetal existence is inscribed, nor can a global preoccupation with the collectivities of plants overlook the singularity of each plant that repeatedly expresses, in a non-exemplary and unrepeatable fashion, vegetal being as such. That is why the loss of a single plant is tantamount to the passing of an entire world.

The fifth offshoot: To respect vegetal being, one must respect the time of plants. Where vegetal life is not spatially ravaged, as it is in the case of deforestation, it is cultivated in a way that infringes on the time of plants. To be sure, vegetal hetero-temporality indifferently invites the time of any other to stand in for the time of the plants themselves. The temporality of capital, nevertheless, violates this botanical "hospitality" itself, in that it imposes the routine of the same—the exigencies of commodification and ever-accelerated profiteering—on crops grown under the auspices of the capitalist agro-scientific complex. Conversely, the necessary precondition for a respectful attitude to vegetal heterotemporality is an approach that leaves the locus of the plant's other vacant and refrains from determining this place once and for all. Not only does the refusal to determine the plants' other extend respect to their being, heteronomously defined by alterity, but it also defers to the environmental conditions and the milieu of the plants' flourishing. Our appreciation of the temporal dimension of vegetal being thus turns into the conduit of a broader environmental ethics.

The sixth offshoot: To harness plants to a particular end is to drive them to ontological exhaustion. On the one hand, the open-endedness, or the essential incompletion, of growth and of other vegetal functions subtracts plants from the logic of actualization they have been often called upon to illustrate. On the other hand, the infinite possibilities of vegetal life correspond to its countless ends: whether channeling inorganic nature and welcoming the other, or furnishing a shared, albeit inappropriable, stratum of life and non-teleologically anticipating the other forms of life, forever anchored in the principles of plant-soul . . . Those, like Hegel, who assert that vegetal beings attain their highest fulfillment in serving as sources of food for animals and humans confine plants to one external end, willfully ignoring their otherwise inexhaustible possibilities. In addition to curtailing vegetal freedom, this theoretical brutality, subtending practical-economic violence, drastically narrows the ontological scope of vegetal life and, in the first instance, the relational ontology binding this life to its human counterpart. For example, the consumption of plants (as sources of nourishment, construction materials, fuel, or even as aesthetic objects) should be recognized as one among many possibilities for our interaction with them and for what comes to define their ends. Plant-thinking does not oppose the use of fruit, roots, and leaves for human nourishment; rather, what it objects to is the total and indiscriminate approach to plants as materials for human consumption within the deplorable framework of the commodified production of vegetal life. If one is to respect vegetal existence, one will facilitate, not restrict, the proliferation of its various ends, not to mention celebrate the lack thereof.

The seventh offshoot: Any ethical guidelines for the consumption of plants should be consistent with the lessons of plant-thinking. When my vegetarian or vegan friends and colleagues ask me, "What can I ethically eat, given your research on plants?" they expect to hear in response that their diet would be incompatible with the full appreciation of vegetal life. And when my non-vegetarian acquaintances rejoice in the idea that those who refuse to eat meat might not be righteous after all, they implicitly come to the same conclusion. We should note from the outset, however, that the question "What can I eat?" is itself wrapped in a set of concerns central to vegetal life with its hallmark capacity for nourishment. Perhaps then it is this very life that holds the answer to our legitimate anxiety about ethical nutrition. I have already noted that plant-thinking does not condemn the consumption of plants and their parts, unless in utilizing them we dim down and disrespect the other facets of ontophytology. This means that the question ought to be reformulated; instead of "*What* can I eat?" we should inquire, "*How* am I to eat ethically?" To put it succinctly, if you wish to eat ethically, *eat like a plant!* Eating like a plant does not entail consuming only inorganic minerals but welcoming the other, forming a rhizome with it, and turning oneself into the passage for the other without violating or dominating it, without endeavoring to swallow up its very otherness in one's corporeal and psychic interiority. The plasticity of vegetal life and its remarkable capacity for regeneration should assist us in following these guidelines. Because plants do not possess an essential core and, moreover, because one of their temporal modalities is "iterability," one can devise an eating pattern that is consonant with their ontology and that does not homogenize them, that does not handle them as instances of the same—as convenient and healthful storehouses of calories, carbohydrates, or other units of stored energy. While the significance of individual decisions and initiatives cannot be overestimated, when it comes to the ethics of eating what is required is a complete and concerted decommodification of vegetal life, a refusal to regulate the human relation to plants on the basis of commodity-economic logic. From genetically modified seeds that do not yield renewable crops to artificially induced scarcity, whereby fruit is thrown away or left to rot in order to maintain high market prices, the capitalist agro-scientific complex militates against vegetal ontology and ethics alike.[6] And, on the contrary, we welcome the vegetal other when we avow its otherness, its irreducibility to a source of either food or profit; we enter into a rhizomatic relation with it when we eat locally grown fruit and vegetables, heeding the wisdom of the plant, whose "reach cannot exceed its grasp";[7] we turn into passages for the vegetal other that nourishes us, if we do not endeavor to consume, along with the edible parts of the actual plants, the temporal modalities and possibilities of vegetal life.

The eighth offshoot: The plant's absolute silence puts it in the position of the subaltern. The absence of voice in plants does not preclude their spatial, material self-expression, though it does pose additional hurdles to the ethical treatment of vegetal life. Our incapacity to communicate with plants the way we do with other human beings (and even with certain animals) by means of a vocal interaction runs the risk of objectifying them or, at best, speaking *for* them, in their defense, if not in their place. In other words, an ethical approach to vegetal beings must tackle the difficult problem of representing them both in the theoretical sense of representation faithful to their ontology and in the strictly political sense of representation as delegation, claiming the right to speak on behalf of the plants that remain immersed in absolute silence. Taken in its first sense, an ethical "representing" awakens the hermeneutics of vegetal existence, circling back to the ontology of vegetation and approximating its living *self-interpretation*, all the while keeping in mind the inherent limits of this endeavor, the perpetual withdrawal of that which is interpreted from the hermeneutical grasp. The second, purely political sense of representation will maintain an ethical edge, on the condition that the "delegates" of plant life, advocating on its behalf, are attuned to the groundlessness of their authority, which has not been bestowed upon them by those they claim to represent. The thresholds of interpretation and delegation must be clearly demarcated in order to ensure that the absolute silence of vegetal life is still respected. How, for instance, could one ethically regret the fading of flowers, if not, as Rilke does,[8] in the language of poetry, which does not represent anything and which, itself, verges on silence?

The ninth offshoot: Rather than sentience, it is the finitude of a living being that furnishes the "yardstick" for ethical treatment. From Schelling to Bergson, philosophers of nature have emphasized that consciousness is not an isolated phenomenon but the other end of the continuum stretching back to the non-conscious life of plants. We may trace this fruitful intuition to the Aristotelian vegetal soul, which, despite its non-sentience, falls under the heading of *psukhē* as the shared stratum of life, thanks to which all other entities are alive. The non-exceptional character of consciousness and sentience complicates the arguments for according a special ethical status to these forms of life at the expense of the vegetal "minimal irritability" (Schelling) and "consciousness asleep" (Bergson). Now, if vegetal life is coextensive with a non-conscious thinking, then an ethical approach to plants is not exhausted by the "moral feeling" of compassion or obligation toward them but is also conceivable as a full-fledged relation, however asymmetrical, which makes this approach genuinely ethical. In Martin Buber's ethical thought, the tree ceases to be a mere "It," objectively known by the science that dissolves it into chemic components, and instead participates in the *I–Thou* relation (*Beziehung*).[9] What sparks off this shift of attitude toward the tree, the shift left unexplained in Buber's philosophy? Does it not attest to the non-verbal and non-conscious communication of one life with another, the reaching of one life's logic over to another? This interaction of heterogeneous "reasons" is ethical on the condition that neither of the two dominates the other but lets the other follow its course, lets it thrive in the manner appropriate to it, lets it be.

The tenth offshoot: The essential incompletion of vegetal life conditions the growth of plant-thinking and ethical action. The "bad infinity" of vegetal proliferation, the continuous striving of the plant to its other without return to itself, the as-yet unexplored possibilities of the vegetal soul and its countless permutations in all living beings—these are but a few indications of the essential incompletion (and, hence, the vivacity, the unrest, and the non-identity) of a life compatible with plant-thinking. Such thinking will not freeze in a doctrinaire form, so long as it keeps its affinity to the ontology of vegetation. Its ethical offshoots, too, exceed the scope of a program of action, or of a fixed set of normative de-contextualized guidelines. Those who are ready to practice plant-thinking must be patient enough to see through the germination of a new ethics from vegetal existence itself, an ethics singularly adapted to each situation, rid of final conclusions, and in tune with our ongoing learning from plants.

Notes

Introduction

1. "I shall, then, take the root of the disputation from plants, which are fixed in the ground by their roots" (Adelard of Bath, *Conversations with His Nephew: Questions on Natural Science*, 93).
2. The recent philosophical literature on this topic is too extensive to allow as much as a representative sampling here. Still, it is worth highlighting the work of the following scholars: Agamben (2003); Buchana (2008); Calarco (2008); Derrida (2008); Francione (2009); Haraway (2007); Linzey (2009); Oliver (2009); Regan (2006); Singer (2002); Sorabji (1995); Steiner (2010); and Wolfe (2003).
3. For the notion of "plant landscape" see Brosse, *L'ordre des choses* (1958).
4. For the lack of human interest in plants, explained with reference to the difference between human and vegetal beings, see Hallé, *In Praise of Plants*, 17ff.
5. Bhattacharya, *Encyclopedia of Indian Philosophies*, 165.
6. Irigaray, *Elemental Passions*, 32.
7. Von Uexküll, *A Foray*, 41.

1. The Soul of the Plant

1. La Mettrie, *Man a Plant*, 86–87.
2. Fichte, *The Science of Rights*, 503. Even for "commonsense" Scottish Enlightenment thinkers this insight was not foreign. According to Thomas Reid, "both vegetables and Animals are United to something immaterial, by such a Union as we conceive between Body and Soul, which Union continues while the Animal or Vegetable is alive, & is dissolved when it dies" (*On the Animate Creation*, 218–219).
3. On the double sense of *arkhē*, see Jacques Derrida's *Archive Fever* (1998). The *Enneads* of Plotinus features an indelible (and anarchic) image of the soul in the shape of the plant: "The Soul of the All (that is, its lowest part) would be like the soul in a great growing plant [*phutō megalo psukhē*], which directs the plant without effort or noise; our lower part would be as if there were maggots in a rotten part of the plant—for that is what the ensouled body [*sōmato empsukhon*] is like in the All" (4.3.4.26–31). The lowest part of the universal soul is a plant that grows by itself, like a weed, without being cultivated; it germinates effortlessly, silently, and, indeed, freely.
4. For more on the non-identity of things, see my *The Event of the Thing* (2009).
5. Hegel, *Philosophy of Nature*, 308.
6. Hegel, *Phenomenology of Spirit*, 436–437.
7. Merleau-Ponty, *Nature*, 3.
8. Cf. Heidegger's *Heraclitus Seminar* (1993). For Heidegger's reading of the Aristotelian equation of *phusis* with Being, see Heidegger, "On the Essence and Concept of *Physis*," as well as Baracchi's insightful essay, "Contributions to the Coming-to-Be of Greek Beginnings" (2006).
9. Heidegger, "On the Essence and Concept of *Physis*," 195.
10. "The theory in the so-called poems of Orpheus presents the same difficulty; for this theory alleges that the soul, borne by the winds, enters into animals when they breathe. Now, this cannot happen to plants, nor to some animals, since they do not all breathe: a point which has escaped those who support this theory" (*De anima* 410b.28–411a3).
11. Hegel, *Philosophy of Nature*, 304.

12. Hegel, *Philosophy of Nature*, 338–339.
13. "While the body desires on its own account . . . , the vegetal soul desires with a desire that stems from something else and through the agency of another" (*Enneads* 4.4.20.22–36).
14. Cf. Stone's *Petrified Intelligence* (2005).
15. Hegel, *Philosophy of Nature*, 346.
16. I am using the concept of "weakness" in keeping with Gianni Vattimo's ideas of "weak thought" and the "weakening of metaphysics."
17. Quoted in Müller-Sievers, *Self-Generation*, 158.
18. Merleau-Ponty, *Nature*, 3.
19. Bergson, *Creative Evolution*, 92. We may recall, in this context, the words of the Tiger-lily in Carroll's *Through the Looking Glass:* "In most gardens," the Tiger-lily said, "they make the beds too soft—so that the flowers are always asleep."
20. Nietzsche, *Will to Power*, 347.
21. For a recent reading of these aspects of *Timaeus*, see Carpenter, "Embodied Intelligent (?) Souls" (2010).
22. "Similarly, Plato averred that plants must know desire, because of the extreme demands of their nutritive capacity. If this were established, it would be in accord with it that they should really know pleasure and pain, and that they should feel. And once this is established, it will be in accord with it that plants should know desire" (*De plantis* 815a22–26).
23. Nietzsche, *Will to Power*, 341–342.
24. Nietzsche, *Will to Power*, 349.
25. Nietzsche, *Will to Power*, 375.
26. Hegel, *Philosophy of Nature*, 304.
27. Nietzsche, *Will to Power*, 367.
28. Nietzsche, *Will to Power*, 374.
29. Cf. Derrida's *The Animal That Therefore I Am* (2008).
30. Marder, *The Event of the Thing*, 29. As for the dispersion of human intentionality, a similar de-idealization befell the unity of Husserl's "consciousness of" and "that of which it is conscious" when Heidegger conceptualized it as a practical and engaged "being-in-the-world," no longer oriented toward a single noematic target, but dispersed in countless modes of concern.

31. Hegel, *Philosophy of Nature*, 303.
32. Fichte conceives of animal soul as "a system of plant-souls" (505). Whereas the soul of a plant is a middle point (*Mittelpunkt*), a crossroads in the process of chemical attraction and repulsion, animal soul, as a combination of multiple plant-souls from which it will never fully distance itself, is a de-centered entity, one where, as Fichte puts it, "every possible point, involving, as it does, a peculiar principle of motion, is the central point of a plant-atmosphere as its lower world" (504–505). The psyche of an animal is a composite of plant-souls that have been de-centered and released to a higher state of freedom, so that the possibility of movement comes to permeate every single part of this synthetic unit. The soul of the animal does not discard its vegetal shape but systematizes and coordinates a multiplicity of plant-souls, from which it is not at all separate, imposing a new finality upon them.
33. On the "extended psychic thing" in psychoanalysis, see my *The Event of the Thing*.
34. Hegel, *Phenomenology of Spirit*, 420.
35. Nietzsche, *Will to Power*, 341.
36. Nietzsche, *Will to Power*, 342 (my emphasis).
37. On the figure of the "innocent plant" in German philosophy, see Miller's *Vegetative Soul* (2002).
38. Nietzsche, *Will to Power*, 328.
39. Compare this to the words of Diotima, as they are related by Socrates in Plato's *Symposium*: "Every mortal thing is preserved in this way; not by keeping it exactly the same forever, like the divine, but by replacing what goes off or is antiquated with something other [*heteron*], in the semblance of the original. Through this device, Socrates, a mortal thing partakes of [*metexei*] immortality" (208a–b). The other way to partake of immortality open to humans is to give birth to beautiful works, ideas, and institutions. One wonders, however, whether this is but a mutation in the vegetal soul responsible, among other things, for procreation. I will take up this question, typical of plant-*thinking*, in chap. 5.
40. Hegel, *Philosophy of Nature*, 304.
41. Hegel, *Phenomenology of Spirit*, 420.

42. The ethical and political effects of plant-thinking will be systematically explored in the sequel to the present volume, titled "Plant-Doing: The Ethics and Politics of Vegetal Life" (manuscript currently in preparation).

2. The Body of the Plant

1. "Metaphysics? What metaphysics do those trees have?" (Fernando Pessoa, poem 5 of *The Keeper of Sheep*).
2. See my "Retracing Capital" (2004).
3. For a representative example of this trend see Matthew Hall's recent *Plants as Persons* (2011).
4. Derrida, *Glas*, 15.
5. As Parkes concludes in *Composing the Soul*, "In view of Nietzsche's fondness for vegetal metaphors, Plato's image of the inverted plant must be anathema: the tree of life *turned upside down!*" (179). In the early modern period this inversion resonates with the refinement of the vegetal metaphorization of metaphysics, for example, in Descartes's famous letter to Picot, the French translator of *Principia philosophiae*, where the philosopher asserts that "the whole of philosophy is like a tree, whose roots are metaphysics, whose trunk is physics, and whose branches, emerging from the trunk, are all the other sciences" (Descartes, *Principles of Philosophy*, xxiv; cf. also Ariew, "Descartes and the Tree of Knowledge" [1992]). The tree of knowledge captured in this description does not merely distill metaphysics to the most vital part of the epistemic plant but also anchors it in a new *topos ouranios* of Cartesian "first philosophy." While the root of this tree is liberated from the darkness of the ground, thanks to the fact that metaphysics necessitates clear and distinct ideas, the scientific branches point downward, to the empirical realities that are the objects of their investigation.
6. Nietzsche, *Writings from the Early Notebooks*, 138.
7. In this respect, Derrida's psychoanalytic reading of Hegel in *Glas* loses its cryptic veneer. Deconstruction's reiterated insight is that the act of castration—or its fetishistic substitute, circumcision—often endows the very phallocentric logic it desires to disband with an ever-greater symbolic

power that finds its exemplary expression in the "possibility of turning upside down [a tree that, nevertheless, continues to thrive, turning branches into new roots,] of the upside-down erection, . . . inscribed in the cycle of the family standing up. The son is son only in his ability to become father, his ability to supply or relieve the father" (81).

The multiple instantiations and institutionalizations of metaphysics, including the allegorical socio-theological incarnations of the inverted tree—the patriarchal genealogy and the divine economy of the Trinity alluded to here—are perpetuated in spite of (and also thanks to) such inversions. To break the spell of metaphysics, it behooves us to put forth an alternative to a simple regrounding of the heavenly plant in the earth, which is the material origin metaphysics has disavowed and debased, and therefore to bracket for the time being the conclusions of the Feuerbachian critique of religion and, by implication, of German idealism. It is advisable instead to resist the temptation to romanticize the plant as the true source—of life, of existence, of being—or as something through which we can reconnect with the "natural origins" of which metaphysics has robbed us.

8. Quoted in Heidegger, *Discourse on Thinking*, 47.
9. Heidegger, *Discourse on Thinking*, 47–48.
10. La Mettrie, *Man a Plant*, 78.
11. Novalis, *Fragmentos*, 133.
12. Oken, *Elements of Physiophilosophy*, 269. Hegel objects to Oken's as much as to Schelling's analogies in his *Philosophy of Nature*.
13. Onfray, *Les formes du temps*, 38.
14. Ponge, *Nouveau Nouveau recueil*, 106.
15. Nancy, *Corpus*, 13.
16. Ponge, *Nouveau Nouveau recueil*, 109.
17. Ponge, *Nouveau Nouveau recueil*, 109.
18. Deleuze and Guattari, *A Thousand Plateaus*, 21.
19. Schelling, *First Outline*, 47.
20. "In nature, the plant alone . . . is vertical, along with man [*L'arbre seul, dans la nature . . . est vertical, avec l'homme*]" (Claudel, *La connaissance de l'Est*, 148).
21. Bachelard, *L'air et les songes*, 266.

22. Bergson, *Creative Evolution*, 102.
23. Onfray, *Les formes du temps*, 26.
24. This interference is due to the fact that the plant is not one and hence is not a totality. When it comes to the theologicial problem of the Trinity, however, Hegel resolves it with the help of the living image of a tree, which turns "what is a contradiction [*Widerspruch*] in the realm of the dead [*im Reiche des Toten*]" into the truth of "the realm of life." The tree's non-individuation means that each of its parts is on a par with the "whole" and that each part—for example, a branch—constitutes another tree within the tree. "A tree which has three branches," Hegel continues, "makes up with them *one* tree [einen *Baum*]; but every 'son' of the tree, every branch (and also its other 'children,' leaves and blossoms) is itself a tree. . . . And it is just as true to say that there is only *one* tree here as to say that there are three" (Hegel, *Early Theological Writings*, 261). A living tree is and is not *one*, precisely because its unity is produced without the construction of an organismic and systemic whole; the obverse of its inseparability from the milieu wherein it grows is its seemingly infinite internal divisibility, permitting one to produce new trees by means of grafting, as Kant had already realized in *Opus postumum*: "Plants permit grafts, and hence aggregates without a system" (182–183). Of course, having forgotten that the "father" tree is not a totalized whole, Hegel runs into at least two internal contradictions. First, contrary to the Christian idea of Trinity, there is no organically conceptual relation either among the transplanted "sons" of the tree, or between these "children" and the "father" tree whence they were derived; second, there is no inherent limit to the number of "sons" that may be obtained from the original tree, if leaves and blossoms, too, were to count as potential trees. The possibility of anarchy and polytheism thus comes to haunt the Hegelian allegory.
25. Canguilhem, *Knowledge of Life*, 113–114. This definition is of course hugely indebted to Jacob von Uexküll, who writes: "The biologist . . . takes into account that each and every living thing is a subject that lives in its own world, of which it is the center" (*A Foray*, 45).
26. Hegel, *Philosophy of Nature*, 323–324.
27. Onfray, *Les formes du temps*, 25.

28. Bachelard outlines the contours of this vegetal articulation. "A straight tree," he writes, "is an obvious force that carries earthly life to the blue sky. . . . The tree gathers and arranges the most diverse elements [*L'arbre droit est une force évident qui porte une vie terrestre au ciel bleu. . . . L'arbre réunit et ordonne les éléments les plus divers*]" (*L'air et les songes*, 263).
29. Or, again: "The tree unites the infernal and the celestial, air and earth [*L'arbre unit l'infernal au céleste, l'air à la terre*]" (Bachelard, *L'air et les songes*, 271).
30. Derrida, *Glas*, 17.
31. The plant, for the author of *De plantis*, is literally rooted outside of itself: "But all herbs, whether they grow above the earth or in it, depend on one of these five conditions: seed, moisture from water, a suitable soil, air, and planting. These five one might say are the roots of plants [*rizai phutōn*]" (827a2–7). Centuries later, La Mettrie would put this dependence in even more derogatory terms: "An appropriate image of a plant is an infant clinging to its nurse's nipple, sucking incessantly. Plants are sucklings of the earth, and they leave the breast only when they die" (*Man a Plant*, 85).
32. Miller, *Vegetative Soul*, 17.
33. Hegel, *Philosophy of Nature*, 307.
34. Hegel, *Philosophy of Nature*, 306.
35. "Unceasingly, unwillingly, we have been carried along by the movement which brings the sun to turn in metaphor; or have been attracted by what turned the philosophical metaphor to the sun. Is this flower of rhetoric (like) a sunflower? That is—but this is not exactly a synonym—analogous to the heliotrope?" (Derrida, *Margins of Philosophy*, 250).
36. Bergson, *Creative Evolution*, 93.
37. Hegel, *Philosophy of Nature*, 309.
38. For such an elaboration, cf. chap. 3 of this study.
39. Hegel, *Philosophy of Nature*, 308.
40. Ponge, *Selected Poems*, 68–69.
41. See Karban, "Plant Behavior," and Karban and Shiojiri, "Self-Recognition Affects Plant Communication."
42. Ponge, *Selected Poems*, 70–71.
43. Husserl, *Ideas I*, 212.

44. In the French philosophical context, Rousseau clearly espoused radical empiricism in relation to plants singularized in every encounter, when he wrote that "with each new blade of grass I encounter, I say to myself with satisfaction: here is yet another plant" (Rousseau, "Reveries of the Solitary Walker," 58). The term "plant," for Rousseau, is semantically rich, to the point of being oversaturated with meaning, because every blade of grass is a new plant altogether, not a slightly different variation on a familiar theme, an imperfect instantiation of the ideal genus.
45. Derrida, *Margins of Philosophy*, 249.
46. Derrida, *The Beast and the Sovereign*, 147ff.
47. Agamben, *Potentialities*, 231.
48. Pessoa, *Obra poetica*, 206.
49. Pessoa, *Obra poetica*, 207.
50. Goethe, *Metamorphosis of Plants*, 65.
51. "Yet [the body] is a skin, variously folded, refolded, unfolded, multiplied, invaginated" (Nancy, *Corpus*, 15).
52. Ovid, *Metamorphoses* 1.9.
53. Goethe, *Metamorphosis of Plants*, 67.
54. Goethe, *Metamorphosis of Plants*, 6 (emphasis added).
55. Goethe, *Metamorphosis of Plants*, 56.
56. Goethe, *Italian Journey*, 310.
57. The conceptualization of the leaf in Goethe and his betrayal of Leibniz stand in the background of Nietzsche's "On Truth and Lies in a Nonmoral Sense": "Every concept arises from the equation of unequal things. Just as it is certain that one leaf is never totally the same as another, so it is certain that the concept 'leaf' is formed by arbitrarily discarding these individual differences. . . . This awakens the idea that, in addition to the leaves, there exists in nature the 'leaf': the original model according to which all the leaves were perhaps woven, sketched, measured, colored, curled, and painted—but by incompetent hands" (*Nietzsche Reader*, 117).
58. Hegel, *Philosophy of Nature*, 303–304.
59. The rejection of the dialectical interplay between sameness and otherness in vegetal life does not preclude its relation to alterity that is not a mere "negation of the same."

60. Quoted in Canguilhem, *Knowledge of Life*, 41.
61. Deleuze and Guattari, A *Thousand Plateaus*, 157.
62. Deleuze and Guattari, A *Thousand Plateaus*, 12.
63. La Mettrie, *Man a Plant*, 89.
64. Goethe, *Metamorphosis of Plants*, 6.
65. Hegel, *Philosophy of Nature*, 344.
66. Hegel, *Philosophy of Nature*, 343.
67. Hegel, *Philosophy of Nature*, 348.
68. Janeczko and Skoczowski, "Mammalian Sex Hormones in Plants," 75.
69. Derrida, "Sexual Difference, Ontological Difference," 72.
70. It is as erroneous to deny the existence of sexuality in plants as to pigeonhole them in one sexual mold—namely the feminine. Elain Miller's *Vegetative Soul* is an attestation of this.
71. Hegel, *Philosophy of Nature*, 323.
72. Derrida, *Dissemination*, 304.
73. Nancy, *Being Singular Plural* (2000).
74. Deleuze and Guattari, A *Thousand Plateaus*, 19.

3. The Time of Plants

1. "The plant faithfully keeps memories of happy daydreams. Every spring, it gives them a new birth" (Gaston Bachelard, *Air and Dreams*; my translation).
2. Cf. Derrida, "*Ousia* and *Grammé*: Note on a Note from *Being and Time*," in *Margins of Philosophy*, 29–67.
3. Hegel, *Logic*, 231–233.
4. Hegel, *Philosophy of Nature*, 323–324.
5. Hegel, *Phenomenology of Spirit*, 2.
6. Heidegger, *Being and Time*, 63.
7. Heidegger, *Being and Time*, 288.
8. Heidegger, *Being and Time*, 287–288.
9. Onfray, *Les formes du temps*, 34.
10. Bergson, *Matter and Memory*, 205, 207.

11. Ponge, *Selected Poems*, 75. Time-lapse photography is another strategy for detecting the time of the plant. For a more meticulous botanical description of this phenomenon, see Hallé, *In Praise of Plants*, 103–104.
12. Nietzsche, *Human, All Too Human*, 114–115.
13. On the question of culture as the afterlife of vegetal existence, two emblematic examples, bread and wine—examples imbued with tremendous religious significance—will demonstrate that the decay of "natural" life is itself harnessed for the purposes of cultural productivity: the fermentation of grapes and of wheat flour is, literally, the afterlife of these respective plants. Hegel is not far beyond our horizon here, and neither is Derrida, who notes a close affinity, indeed a near sameness, of "spirit (*Geest*, *Geist*) and fermentation (*gäschen*)" (*Glas*, 59).
14. Levinas, "Place and Utopia," *Difficult Freedom*, 100.
15. Levinas, *Difficult Freedom*, 100.
16. Levinas, *Time and the Other*, 79.
17. Levinas, *Otherwise than Being*, 15.
18. Bachelard, *L'air et les songes*, 283.
19. Hegel, *Philosophy of Nature*, 342.
20. Novalis, *Fragmentos*, 123.
21. Nietzsche, *Will to Power*, 349.
22. Novalis, *Fragmentos*, 84.
23. Hegel, *Philosophy of Nature*, 305, 308. The philosophical consensus on the subject of plant nutrition has affected even a thinker as sophisticated as Bergson, who notes that "vegetables continually and mechanically draw these [nutritive] elements from an environment that continually provides it. Animals, by action that is discontinuous, concentrated in certain moments, and conscious, go to find these bodies in organisms that have already fixed them" (*Creative Evolution*, 93).
24. "Seasonal existence [is . . .] opposed to a notion of time delineated by 'a beginning' and 'an end'" (Jullien, *Vital Nourishment*, 7).
25. Levinas, *Totality and Infinity*, 34.
26. Goethe, *Metamorphosis of Plants*, 100.
27. Nancy, *Muses*, 24.
28. Derrida, *Limited Inc.*, 59.

29. Hegel, *Philosophy of Nature*, 316.
30. Hegel, *Philosophy of Nature*, 342.
31. Levinas, *Totality and Infinity*, 268.
32. Bergson, *Creative Evolution*, 14.
33. Heidegger, *Fundamental Concepts of Metaphysics*, 25.
34. Onfray, *Les formes du temps*, 26.
35. Onfray, *Les formes du temps*, 30.
36. Deleuze, *Difference and Repetition*, 1.
37. Deleuze, *Difference and Repetition*, 8.
38. Ponge, *Selected Poems*, 71.
39. Deleuze, *Difference and Repetition*, 53.
40. "The question is asked why Nature repeats: because it is *partes extra partes*, *mens momentanea*" (Deleuze, *Difference and Repetition*, 16).
41. It matters little whether the plant is perennial or not, above all because the repetition of a leaf is a feature of annual plants as well.
42. Cf. Pesic, *Seeing Double*, 54ff.
43. Cf. Hegel's *Logic*, paragraph 117: "Diversity."
44. Derrida, *Limited Inc.*, 119.
45. Bachelard, *L'air et les songes*, 262.

4. The Freedom of Plants

1. "To free ourselves, let us free the flower" (Francis Ponge, [New "New collection, 1967–1984"]).
2. Nancy, *The Experience of Freedom*, 1.
3. "The term *freedom*, which supposes a puerile or oratorical enthusiasm, is from the outset fallacious" (Bataille, *Oeuvres complètes*, 131).
4. "The name *natura* first of all signifies the uterine orifice, the place where birth occurs. The natural is what follows from being-through-birth—from natality" (Schürmann, *Broken Hegemonies*, 200).
5. Aristotle, *Physics* 192b21–23.
6. Cf. Aristotle, *Physics* 199b15ff.
7. Derrida, *Glas*, 86.
8. Derrida, *Truth in Painting*, 92.

9. Derrida, *Truth in Painting*, 89.
10. Derrida, *Glas*, 13.
11. Derrida, *Glas*, 12.
12. Hegel, *Philosophy of Nature*, 307.
13. Hegel, *Philosophy of Nature*, 320.
14. Hegel, *Aesthetics*, 137.
15. Hegel, *Aesthetics*, 247.
16. Kant, *Critique of Judgement*, 88.
17. Ponge, *Nouveau Nouveau recueil*, 104.
18. Ponge, *Selected Poems*, 79 (translation modified).
19. "The *shape* of the plant, as not liberated out of individuality into subjectivity, is still closely related to geometrical forms and crystalline regularity" (Hegel, *Philosophy of Nature*, 311).
20. Bergson, *Creative Evolution*, 89.
21. Heidegger, *Basic Writings*, 230.
22. While Jacob von Uexküll performed an invaluable philosophical service in his quasi-phenomenological descriptions of animal worlds, he did not entertain the possibility of plant worlds but rather relegated plants to the background of animal *Umwelten*. Yet even his famous example of the blind and deaf tick, with a world of its own, highlights the predominance of vegetal functions—photosensitivity and thermosensitivity—in this animal (*Foray*, 44–45).
23. Cf. Heidegger, *Fundamental Concepts of Metaphysics*, 236ff.
24. Heidegger, *Basic Concepts*, 4.
25. Cf. the epigraph to the present chapter.
26. Grondin, *Du sens de la vie*, 60.
27. Hegel, *Philosophy of Nature*, 349.
28. La Mettrie, *Man a Plant*, 85.
29. Ponge, *Selected Poems*, 69.
30. Even so staunch a proponent of a certain vegetal animism as Plotinus concedes that the desire of trees is, so to speak, de-sensitized, devoid of anger or any strong emotions, because they can register but a vague feeling of irritation (3.4.28.58–63).
31. Cf. Derrida, *Of Spirit*, 19.
32. McGinnis, *Avicenna*, 262.

33. Ponge, *Nouveau Nouveau recueil*, 127.
34. Nietzsche, *Will to Power*, 219.
35. Nietzsche, *Will to Power*, 235.
36. Marx, *Early Political Writings*, 64.
37. Heidegger, *Fundamental Concepts of Metaphysics*, 171.
38. Heidegger, *Fundamental Concepts of Metaphysics*, 77.
39. Needless to say, the cleansing, bespeaking the freedom of nihilism, should not be conflated with the ideal of metaphysical purity, in that, in contrast to the latter, it hinges on a necessary cross-contamination of the plant and the human.
40. Kant, *Critique of Practical Reason*, 17.
41. Kant, *Critique of Practical Reason*, 53.
42. Levinas, *Otherwise than Being*, 4.
43. Levinas, *God, Death, and Time*, 175.
44. Levinas, *Totality and Infinity*, 134.
45. Kant, *Critique of Judgement*, 72.
46. Kant, *Critique of Judgement*, 76.
47. Kant, *Critique of Judgement*, 72.
48. Nietzsche, *Writings from the Early Notebooks*, 36.
49. Kant, *Critique of Judgement*, 72–73.
50. Kant, *Critique of Judgement*, 43.
51. Ponge, *Selected Poems*, 69.
52. Derrida, *The Truth in Painting*, 95.
53. Pollan, *The Botany of Desire*, 10.
54. Schiller, *On the Aesthetic Education of Man*, 133.
55. Schiller, *On the Aesthetic Education of Man*, 134.
56. Schiller, *On the Aesthetic Education of Man*, 134.
57. Schiller, *On the Aesthetic Education of Man*, 135.
58. Deleuze and Guattari, *A Thousand Plateaus*, 11.
59. Hegel, *Phenomenology of Spirit*, 420.
60. Hegel, *Phenomenology of Spirit*, 420.
61. Horkheimer and Adorno, *Dialectic of Enlightenment*, 49.
62. "Thence we sailed on, grieved at heart, and we came to the land of the Cyclopes, an overweening and lawless folk, who, trusting in the immortal

gods, plant nothing with their hands nor plough; but all these things spring up for them without sowing or ploughing, wheat, and barley, and vines, which bear the rich clusters of wine, and the rain of Zeus gives them increase" (*Odyssey* 9.105–115).

63. Derrida, *Glas*, 15.
64. Derrida, *Glas*, 20.
65. Pollan, *The Botany of Desire*, 66.
66. Marx, *Early Political Writings*, 58.

5. The Wisdom of Plants

1. "Flora teaches its truths to the one who knows how to listen to it, and then to understand it" (Michel Onfray, [The forms of time]).
2. Nietzsche, *Will to Power*, 349.
3. Levinas, *Entre nous*, 128.
4. Bataille, *Theory of Religion*, 19.
5. "Can we go further and say that life, like conscious activity, is invention, is unceasing creation?" (Bergson, *Creative Evolution*, 19).
6. Bergson, *Creative Evolution*, 117.
7. Nietzsche, *Writings from the Early Notebooks*, 139, 140.
8. Cf. Viner et al., "Ca^{2+} and Phytochrome Control of Leaf Unrolling."
9. Cf. Verdus et al., "Storage of Environmental Signals in Flax."
10. Schelling, *First Outline*, 146.
11. Schelling, *First Outline*, 182–183.
12. For a much more recent example, refer to the thesis regarding the unity of consciousness of all living beings advanced, among others, by Raoul Heinrich Francé in *Germs of Mind in Plants*: "Even if all our hopes are not realized, we have brought away a mighty knowledge that reaches down into the very depths of all being: *the certainty that the life of plants is one with that of animals, and with that of ourselves*" (Francé, *Germs of Mind*, 139).
13. Along these lines, Samuel Butler writes, with a great deal of irony, in *Erewhon*: "We say that the oak and the rose are unintelligent, and on finding that they do not understand our business conclude that they do not

understand their own. But what can a creature who talks in this way know about intelligence? Which shows greater signs of intelligence? He, or the rose and oak?" (237). See also the celebrated work of Jakob von Uexküll, *A Foray Into the Worlds of Animals and Humans* (2010).

14. Cf. Fechner, *Nanna*, 53: "In the ray of the sun [the plant] could still gain a feeling that it is elevated above its former sphere as we are by receiving the divine in our souls."
15. Hegel, *Philosophy of Nature*, 306 (emphasis added). A similar observation on the conduct of potatoes is to be found in the writings of Samuel Butler: "Even a potato in a dark cellar has a certain low cunning about him which serves him in excellent stead. He knows perfectly well what he wants and how to get it. He sees the light coming from the cellar window and sends his shoots crawling straight thereto: they will crawl along the floor and up the wall and out the cellar window; if there be a little earth anywhere on the journey he will find it and use it for his own ends" (*Erewhon*, 200).
16. Hegel, *Philosophy of Nature*, 306.
17. Cf. Takahashi and Scott, "Intensity of Hydrostimulation."
18. Cf. Aphalo and Ballare, "On the Importance of Information-Acquiring Systems in Plant–Plant Interactions."
19. Heidegger, *Zollikon Seminars*, 217.
20. Merleau-Ponty, *Phenomenology of Perception*, 140, n. 54.
21. Nietzsche, *Human, All Too Human*, 21.
22. Nietzsche, *Human, All Too Human*, 21.
23. Nietzsche, *Will to Power*, 273.
24. Deleuze and Guattari, *A Thousand Plateaus*, 11.
25. Karen Houle has taken some of the first steps along this path in her recent article, "Animal, Vegetable, Mineral."
26. Bergson, *Creative Evolution*, xx.
27. Bergson, *Creative Evolution*, xix.
28. Bergson, *Creative Evolution*, xix.
29. Heidegger, "The Anaximander Fragment," 42.
30. Bateson, *Steps to an Ecology of Mind*, 491.
31. Bateson, *Steps to an Ecology of Mind*, 492.
32. Pascal, *Pensées*, 66.

33. Deleuze and Guattari, *A Thousand Plateaus*, 11.
34. Deleuze and Guattari, *A Thousand Plateaus*, 15.
35. Nietzsche, *Human, All Too Human*, 113.
36. Novalis, *Fragmentos*, 79.
37. Novalis, *Philosophical Writings*, 102–103.
38. "Despairing of the reality of those things, and completely assured of their nothingness, [the animals] fall to without ceremony [*sic*] and eat them up" (Hegel, *Phenomenology of Spirit*, 65).
39. Derrida et al., "An Interview with Jacques Derrida," 4. A century before Derrida, Nietzsche had observed: "Modern man understands how to digest many things, *indeed almost everything*—it is his kind of ambition" (*Daybreak*, 170 [emphasis added]). This "almost everything" constitutes the material limit to digestive assimilation, the indigestible deconstructive remainder invoked by Derrida.
40. The deconstructive remainder resonates with the non-identical in Theodor Adorno's negative dialectics and with Santiago Zabala's "remains of Being" (2009).
41. The reproductive function does not, of course, exhaust the vast possibilities of human sexuality, least of all in Freud. But no matter however transformed, this sexuality is a variation on the vegetal theme.
42. Freud, "Three Essays on Sexuality," 194 (emphasis added).
43. Freud, "Three Essays on Sexuality," 197.
44. With customary precision, Nietzsche gestures toward this conclusion when he writes, regarding "the possibility of progress," that "men are capable of *consciously* resolving to evolve themselves to a new culture, whereas formerly they did so unconsciously and fortuitously: they can now . . . manage the earth as a whole economically, balance and employ the powers of men in general. This new, conscious culture destroys the old, which viewed as a whole has led an unconscious animal- and plant-life" (*Human, All Too Human*, 25).
45. Novalis, *Fragmentos*, 115.
46. Freud, "Three Essays on Sexuality," 181–182.
47. For an outstanding theoretical account of this transition, see Vieira, *Seeing Politics Otherwise*.

48. Abrams, *The Mirror and the Lamp*, 201ff.
49. Nietzsche, *Will to Power*, 226.
50. Nietzsche, *Human, All Too Human*, 122.
51. Shestov, [The apotheosis of groundlessness], 119 (my translation).

Epilogue

1. "And words? Where do they go? How many of them remain? For how long? And what for, after all? . . . I think of my grandfather Jerónimo, who in his final hours went to bid farewell to the trees he had planted, embracing them and weeping because he wouldn't see them again. It's a lesson worth learning. So I embrace the words I have written, I wish them a long life, and resume my writing where I left off. There can be no other response" (José Saramago, *The Notebook*).
2. See the tree diagram "Moral Considerations of Plants for Their Own Sake" on p. 6 of the Swiss report.
3. Cf. p. 8 of the Swiss report.
4. The ethical "offshoots" presented here in a condensed form will germinate in the sequel to *Plant-Thinking*, which is currently being composed under the tentative title "Plant-Doing: The Ethics and Politics of Vegetal Life."
5. This question will likewise be raised in "Plant-Doing: The Ethics and Politics of Vegetal Life."
6. For an incisive analysis of capitalist agriculture in its current form, see Albritton, *Let Them Eat Junk*.
7. Ponge, *Selected Poems*, 72.
8. Cf. the epigraph to this chapter.
9. Buber, *I and Thou*, 15.

Works Cited

Abrams, M. H. *The Mirror and the Lamp: Romantic Theory and the Critical Tradition*. Oxford: Oxford University Press, 1971.

Adelard of Bath. *Conversations with His Nephew: On the Same and the Different; Questions on Natural Science; and On Birds*. Ed. and trans. Charles Burnett. Cambridge: Cambridge University Press, 1998.

Agamben, Giorgio. *The Open: Man and Animal.* Trans. Kevin Attell. Stanford: Stanford University Press, 2003.

———. *Potentialities: Collected Essays in Philosophy*. Ed. and trans. Daniel Heller-Roazen. Stanford: Stanford University Press, 1999.

Albritton, Robert. *Let Them Eat Junk: How Capitalism Creates Hunger and Obesity*. London: Pluto Press, 2009.

Aphalo, P. J., and C. L. Ballare. "On the Importance of Information-Acquiring Systems in Plant–Plant Interactions." *Functional Ecology* 9 (1995): 5–14.

Aquinas, Saint Thomas. *The Summa Theologica of Saint Thomas Aquinas*. Trans. fathers of the English Dominican Province; rev. Daniel J. Sullivan. Chicago: Encyclopedia Britannica, 1952.

Ariew, Roger. "Descartes and the Tree of Knowledge." *Synthese* 92.1 (1992): 101–116.

Aristotle. *The Athenian Constitution; The Eudemian Ethics; On Virtues and Vice*. Loeb Classical Library, vol. 285. Cambridge: Harvard University Press, 1935.

———. *Metaphysics*. Loeb Classical Library, vols. 271 and 287. Cambridge: Harvard University Press, 1933–1935.

———. *Nicomachean Ethics*. Loeb Classical Library, vol. 19, 2nd ed. Cambridge: Harvard University Press, 1934.

———. *The Physics*. Loeb Classical Library, vols. 228 and 255, rev. ed. Cambridge: Harvard University Press, 1934–1957.

———. "De plantis." In *Minor Works*, 141–236. Loeb Classical Library, vol. 307. Cambridge: Harvard University Press, 1963.

———. *On the Soul; Parva naturalia; On Breath*. Loeb Classical Library, vol. 288, rev. ed. Cambridge: Harvard University Press, 1975.

Bachelard, Gaston. *L'air et les songes: Essai sur l'imagination du mouvement*. Paris: José Corti, 1943.

Baracchi, Claudia. "Contributions to the Coming-to-Be of Greek Beginnings: Heidegger's Inceptive Thinking." In *Heidegger and the Greeks: Interpretive Essays*, ed. Drew Hyland and John Manoussakis, 23–42. Bloomington: Indiana University Press, 2006.

Bataille, Georges. *Oeuvres complètes*, vol. 2. Paris: Le Seuil, 1970.

———. *Theory of Religion*. Trans. Robert Hurley. New York: Zone, 1992.

Bateson, Gregory. *Steps to an Ecology of Mind: Collected Essays in Anthropology, Psychiatry, Evolution, and Epistemology*. Chicago: University of Chicago Press, 2000.

Bergson, Henri. *Creative Evolution*. New York: Barnes and Noble, 2005.

———. *Matter and Memory*. Trans. N. M. Paul and W. S. Palmer. New York: Zone, 1990.

Bernstein, Jay M. *Adorno: Disenchantment and Ethics*. Cambridge: Cambridge University Press, 2001.

Bhattacharya, Sibajiban. *The Encyclopedia of Indian Philosophies*, vol. 10: *Jain Philosophy*. New Delhi: American Institute of Indian Studies, 1970.

Brosse, Jacques. *L'ordre des choses*. Paris: Plon, 1958.

Buber, Martin. *I and Thou*. Trans. Ronald Gregor Smith. London: Continuum, 2004.

Buchana, Brett. *Onto-Ethologies: The Animal Environments of Uexküll, Heidegger, Merleau-Ponty, and Deleuze*. Albany: SUNY Press, 2008.

Butler, Samuel. *Erewhon*. London: Penguin, 1985.

Calarco, Matthew. *Zoographies: The Question of the Animal from Heidegger to Derrida*. New York: Columbia University Press, 2008.

Canguilhem, Georges. *Knowledge of Life*. Trans. Stephanos Geroulantos and Daniela Ginsburg. New York: Fordham University Press, 2008.

Carpenter, Amber. "Embodied Intelligent (?) Souls: Plants in Plato's *Timaeus*." *Phronesis: A Journal of Ancient Philosophy* 55.4 (2010): 281–303.

Claudel, Paul. *La connaissance de l'Est*. Paris: Gallimard, 2000.

Deleuze, Gilles. *Difference and Repetition*. Trans. Paul Patton. New York: Continuum, 2004.

Deleuze, Gilles, and Felix Guattari. *A Thousand Plateaus: Capitalism and Schizophrenia*. Trans. Brian Massumi. Minneapolis: University of Minnesota Press, 1987.

Derrida, Jacques. *The Animal That Therefore I Am*. Trans. David Wills. New York: Fordham University Press, 2008.

———. *Archive Fever: A Freudian Impression*. Trans. Eric Prenowitz. Chicago: University of Chicago Press, 1998.

———. *The Beast and the Sovereign*, vol. 1. Trans. Geoffrey Bennington. Chicago: University of Chicago Press, 2009.

———. *Dissemination*. Trans. Barbara Johnson. Chicago: University of Chicago Press, 1983.

———. *Glas*. Trans. John P. Leavey and Richard Rand. Lincoln: University of Nebraska Press, 1986.

———. *Limited Inc*. Trans. Samuel Weber. Evanston, IL: Northwestern University Press, 1988.

———. *Margins of Philosophy*. Trans. Alan Bass. Chicago: University of Chicago Press, 1985.

———. "Sexual Difference, Ontological Difference." *Research in Phenomenology* 8 (1978): 65–83.

———. *Of Spirit: Heidegger and the Question*. Trans. Geoffrey Bennington. Chicago: University of Chicago Press, 1991.

———. *The Truth in Painting*. Trans. Geoffrey Bennington. Chicago: University of Chicago Press, 1987.

Derrida, Jacques, Daniel Birnbaum, and Anders Olsson. "An Interview with Jacques Derrida on the Limits of Digestion." *E-Flux Journal* 2, Jan. 2009. http://worker01.e-flux.com/pdf/article_33.pdf (accessed Oct. 21, 2010).

Descartes, René. *Principles of Philosophy*. Trans. Valentine Miller. Dodrecht: Kluwer, 1991.

Durkheim, Emile. *The Elementary Forms of Religious Life*. Trans. Karen Fields. New York: Free Press, 1995.

Fechner, Gustav Theodor. *Nanna; oder, Über das Seelenleben der Pflanzen*. 5th ed. Leipzig: Leopold Voß, 1921.

Fichte, J. G. *The Science of Rights*. Trans. A. E. Kroeger. London: Routledge and Kegan Paul, 1970.

Francé, Raoul. *Germs of Mind in Plants*. Trans. A. M. Simons. Chicago: Charles H. Kerr, 1905.

Francione, Gary. *Animals as Persons: Essays on the Abolition of Animal Exploitation*. New York: Columbia University Press, 2009.

Freud, Sigmund. "The Interpretation of Dreams, Part II." In *The Standard Edition of the Complete Psychological Works of Sigmund Freud*, 5:339–630, trans. and ed. James Strachey. London: Vintage, 2001.

———. "Three Essays on Sexuality." In *The Standard Edition of the Complete Psychological Works of Sigmund Freud*, 7:125–244, trans. and ed. James Strachey. London: Vintage, 2001.

———. *Totem and Taboo*. Trans. James Strachey. London: Routledge and Kegan Paul, 1950.

Goethe, Johann Wolfgang. *Italian Journey*. London: Collins/Penguin, 1962.

———. *The Metamorphosis of Plants*. Cambridge: MIT Press, 2009.

Grondin, Jean. *Du sens de la vie*. Montreal: Bellarmin, 2003.

Hall, Matthew. *Plants as Persons: A Philosophical Botany*. Albany: SUNY Press, 2011.

Hallé, Francis. *In Praise of Plants*. Cambridge: Timber Press, 2002.

Haraway, Donna. *When Species Meet*. Minneapolis: University of Minneapolis Press, 2007.

Hegel, G. W. F. *Aesthetics: Lectures on Fine Art*. Trans. T. Knox. Oxford: Oxford University Press, 1998.

———. *Early Theological Writings*. Trans. T. M. Knox. Chicago: University of Chicago Press, 1975.

———. *Hegel's Logic: Being Part One of the Encyclopaedia of the Philosophical Sciences*. Trans. William Wallace. Oxford: Oxford University Press, 1975.

———. *Hegel's Philosophy of Nature: Encyclopaedia of the Philosophical Sciences (1830), Part II*. Trans. A. V. Miller. Oxford: Oxford University Press, 2004.

———. *Phenomenology of Spirit*. Trans. A. V. Miller. Oxford: Oxford University Press, 1979.

Heidegger, Martin. "The Anaximander Fragment." In *Early Greek Thinking*, 13–58. Trans. David Farrell Krell and F. A. Capuzzi. New York: Harper and Row, 1984.

———. *Basic Concepts*. Trans. Gary E. Aylesworth. Bloomington: Indiana University Press, 1998.

———. *Basic Writings*. Ed. David Farrell Krell. Rev. ed. New York: Harper and Row, 1993.

———. *Being and Time*. Trans. John Macquarrie and Edward Robinson. New York: Harper and Row, 1962.

———. *Discourse on Thinking*. Trans. John Anderson and Hans Freund. New York: Harper and Row, 1966.

———. "On the Essence and Concept of *Physis* in Aristotle's *Physics* B, 1." In *Pathmarks*. Trans. W. McNeill. Cambridge: Cambridge University Press, 1998.

———. *The Fundamental Concepts of Metaphysics: World, Finitude, Solitude*. Trans. William McNeill and Nicholas Walker. Bloomington: Indiana University Press, 1995.

———. *Heraclitus Seminar*. Trans. Charles Seibert. Evanston, IL: Northwestern University Press, 1993.

———. *Zollikon Seminars: Protocols–Conversations–Letters*. Ed. Medard Boss. Evanston, IL: Northwestern University Press, 2001.

Homer. *The Odyssey*. Loeb Classical Library, vols. 104–105. Cambridge: Harvard University Press, 1960.

Horkheimer, Max, and Theodor Adorno. *Dialectic of Enlightenment*. Trans. John Cumming. Stanford: Stanford University Press, 2001.

Houle, Karen. "Animal, Vegetable, Mineral: Ethics as Extension or Becoming? The Case of Becoming-Plant." *Journal for Critical Animal Studies* 9 (2011): 89–116.

Husserl, Edmund. *Ideas Pertaining to a Pure Phenomenology and to a Phenomenological Philosophy: First Book*. Trans. F. Kersten. Dodrecht: Kluwer Academic, 1983.

Irigaray, Luce. *Elemental Passions*. Trans. Joanne Collie and Judith Still. London: Athlone Press, 1992.

Janeczko, Anna, and Andrzej Skoczowski. "Mammalian Sex Hormones in Plants." *Folia Histochemica et Cytobiologica* 43.2 (2005): 71–79.

Jonas, Hans. *The Phenomenon of Life: Toward a Philosophical Biology*. Evanston, IL: Northwestern University Press, 2001.

Jullien, François. *Vital Nourishment: Departing from Happiness*. Trans. Arthur Goldhammer. New York: Zone, 2007.

Kant, Immanuel. *Critique of Judgement*. Trans. James C. Meredith. Oxford: Clarendon Press, 1953.

———. *Critique of Practical Reason*. Ed. Mary Gregor. Cambridge: Cambridge University Press, 1997.

———. *Opus postumum*. Ed. Eckart Förster. Cambridge: Cambridge University Press, 1993.

Karban, Richard. "Plant Behavior and Communication." *Ecology Letters* 11.7 (2008): 727–739.

Karban, Richard, and K. Shiojiri. "Self-Recognition Affects Plant Communication and Defense." *Ecology Letters* 12.6 (2009): 502–506.

La Mettrie, Julien Offray de. *Man a Machine; and, Man a Plant*. Trans. Richard Watson and Maya Rybalka. Indianapolis: Hackett, 1994.

Levinas, Emmanuel. *Difficult Freedom: Essays on Judaism*. Trans. Sean Hand. Baltimore: Johns Hopkins University Press, 1997.

———. *Entre Nous: Thinking-of-the-Other*. Trans. Michael B. Smith and Barbara Harshav. New York: Columbia University Press, 1998.

———. *God, Death, and Time*. Trans. Bettina Bergo. Stanford: Stanford University Press, 2000.

———. *Otherwise than Being; or, Beyond Essence*. Trans. Alphonso Lingis. The Hague: Martinus Nijhoff, 1981.

———. *Time and the Other*. Trans. Richard Cohen. Pittsburgh: Duquesne University Press, 1990.

———. *Totality and Infinity: An Essay on Exteriority*. Trans. Alphonso Lingis. Pittsburgh: Duquesne University Press, 1969.

Linzey, Andrew. *Why Animal Suffering Matters: Philosophy, Theology, and Practical Ethics*. Oxford: Oxford University Press, 2009.

Marder, Michael. *The Event of the Thing: Derrida's Post-Deconstructive Realism*. Toronto: University of Toronto Press, 2009.

———. "Retracing Capital: Toward a Theory of Trace in Marxian Political Economy." *Rethinking Marxism* 16.3 (2004): 243–259.

Marx, Karl. *Early Political Writings*. Ed. Joseph O'Malley. Cambridge: Cambridge University Press, 1994.

McGinnis, Jon. *Avicenna*. Oxford: Oxford University Press, 2010.

Merleau-Ponty, Maurice. *Nature: Course Notes from the Collège de France*. Trans. Robert Vallier. Evanston, IL: Northwestern University Press, 2003.

———. *Phenomenology of Perception*. Trans. Colin Smith. London: Routledge, 2002.

Miller, Elaine. *The Vegetative Soul: From Philosophy of Nature to Subjectivity in the Feminine*. Albany: SUNY, 2002.

Muller-Sievers, Helmut. *Self-Generation: Biology, Philosophy, and Literature Around 1800*. Stanford: Stanford University Press, 1997.

Nancy, Jean-Luc. *Being Singular Plural*. Trans. Robert Richardson and Anne O'Byrne. Stanford: Stanford University Press, 2000.

———. *Corpus*. Trans. Richard Rand. New York: Fordham University Press, 2008.

———. *The Experience of Freedom*. Trans. Bridget McDonald. Stanford: Stanford University Press, 1994.

———. *The Muses*. Trans. Peggy Kamuf. Stanford: Stanford University Press, 1997.

Nietzsche, Friedrich. *Daybreak: Thoughts on the Prejudices of Morality*. Trans. Maudemarie Clark and Brian Leiter. Cambridge: Cambridge University Press, 1997.

———. *Human, All Too Human: A Book for Free Spirits*. Trans. R. J. Hollingdale. Cambridge: Cambridge University Press, 1986.

———. *The Nietzsche Reader*. Ed. Keith Ansell Pearson and Duncan Large. Malden, MA: Blackwell, 2006.

———. *The Will to Power*. Trans. Walter Kaufman and R. J. Hollingdale. New York: Vintage, 1968.

———. *Writings from the Early Notebooks*. Ed. Raymond Geuss and Alexander Nehamas. Cambridge: Cambridge University Press, 2009.

Novalis. *Fragmentos de Novalis*. Ed. Rui Chafes. Lisbon: Assirio and Alvim, 1992.

———. *Philosophical Writings*. Ed. and trans. Margaret Mahoney Stoljar. Albany: SUNY Press, 1997.

Oken, Lorenz. *Elements of Physiophilosophy*. Trans. Alfred Tulk. London: Ray Society, 1847.

Oliver, Kelly. *Animal Lessons: How They Teach Us to Be Human*. New York: Columbia University Press, 2009.

Onfray, Michel. *Les formes du temps: Théorie du Sauternes*. Paris: Mollat, 2009.

Ovid. *Metamorphoses*, bks. I–III. Loeb Classical Library, vol. 42. Cambridge: Harvard University Press, 1984.

Parkes, Graham. *Composing the Soul: Reaches of Nietzsche's Psychology*. Chicago: University of Chicago Press, 1994.

Pascal, Blaise. *Pensées*. Trans. A. J. Krailsheimer. London: Penguin, 1995.

Pesic, Peter. *Seeing Double: Shared Identities in Physics, Philosophy, and Literature*. Cambridge: MIT Press, 2003.

Pessoa, Fernando. *Obra Poetica*. Rio de Janeiro: José Aguilar, 1969.

———. *Selected Poems*. Trans. Richard Zenith. New York: Grove Press, 1998.

Plato. *Euthyphro; Apology; Crito; Phaedo; Phaedrus*. Loeb Classical Library, vol. 36. Cambridge: Harvard University Press, 1914.

———. *Lysis; Symposium; Gorgias*. Loeb Classical Library, vol. 166. Cambridge: Harvard University Press, 1925.

———. *The Republic*. Loeb Classical Library, vols. 237 and 276. Cambridge: Harvard University Press, 1930–1935.

———. *Timaeus; Critias; Cleitophon; Menexenus; Epistles*. Loeb Classical Library, vol. 234. Cambridge: Harvard University Press, 1929.

Pollan, Michael. *The Botany of Desire: A Plant's-Eye View of the World*. New York: Random House, 2001.

Plotinus. *The Enneads*. Trans. A. Armstrong. Loeb Classical Library, vols. 440–445, 468. Cambridge: Harvard University Press, 1966–1988.

Ponge, Francis. *Nouveau Nouveau recueil, 1967–1984*. Paris: Gallimard, 1992.

———. *Selected Poems*. Ed. Margaret Guiton. Winston-Salem: Wake Forest University Press, 1994.

Porphyry. *Introduction*. Trans. Jonathan Barnes. Oxford: Oxford University Press, 2006.

Regan, Tom. *Defending Animal Rights*. Urbana: University of Illinois Press, 2006.

Reid, Thomas. *On the Animate Creation: Papers Relating to the Life Sciences*. Ed. Paul Wood. Edinburgh: Edinburgh University Press, 1995.

Rousseau, Jean-Jacques. "Reveries." In *Collected Writings*, vol. 8:1-90, ed. Roger Masters and Christopher Kelly. Hanover, NH: University of New England Press, 2000.

Schelling, F. W. J. *First Outline of a System of the Philosophy of Nature*. Trans. Keith Peterson. Albany: SUNY Press, 2004.

Schiller, Friedrich. *On the Aesthetic Education of Man*. Trans. Reginald Snell. Mineola, NY: Dover, 2004.

Schürmann, Reiner. *Broken Hegemonies*. Trans. Reginald Lilly. Bloomington: Indiana University Press, 2003.

Shelley, P. B. *Shelley on Love: An Anthology*. Ed. Richard Holmes. Berkeley: California University Press, 1980.

Shestov, Lev. *Apofeoz bezpochvennosti/The Apotheosis of Groundlessness*. Moscow: ACT, 2004.

Singer, Peter. *Animal Liberation*. New York: Harper Collins, 2002.

Sorabji, Richard. *Animal Minds and Human Morals: The Origins of the Western Debate*. Ithaca: Cornell University Press, 1995.

Steiner, Gary. *Anthropocentrism and Its Discontents: The Moral Status of Animals in the History of Western Philosophy*. Pittsburgh: University of Pittsburgh Press, 2010.

Stone, Alison. *Petrified Intelligence: Nature in Hegel's Philosophy*. Albany: SUNY Press, 2005.

Swiss Federal Ethics Committee on Non-Human Biotechnology. "The Dignity of Living Beings with Regard to Plants: Moral Consideration of Plants for Their Own Sake." 2008. http://www.ekah.admin.ch/fileadmin/ekah-dateien/dokumentation/publikationen/e-Broschure-Wurde-Pflanze-2008.pdf (accessed on Oct. 21, 2011).

Takahashi, H., and T. K. Scott. "Intensity of Hydrostimulation for the Induction of Root Hydrotropism and Its Sensing by the Root Cap." *Plant, Cell, and Environment* 16 (1993): 99–103.

Theophrastus. *Enquiry Into Plants*, vol. 1, bks. 1–5. Loeb Classical Library, vol. 70. Cambridge, MA: Harvard University Press, 1916.

Verdus, M. C., M. Thellier, and C. Ripoll. "Storage of Environmental Signals in Flax: Their Morphogenetic Effects as Enabled by a Transient Depletion of Calcium." *Plant Journal* 12 (1997): 1399–1410.

Vieira, Patricia. *Seeing Politics Otherwise: Vision in Latin American and Iberian Fiction*. Toronto: University of Toronto Press, 2011.

Viner, N., G. Whitlam, and H. Smith. "Ca^{2+} and Phytochrome Control of Leaf Unrolling in Dark-Grown Barley Seedlings." *Planta* 175 (1988): 209–213.

von Uexküll, Jakob. A Foray Into the Worlds of Animals and Humans. Trans. Joseph D. O'Neill. Minneapolis: University of Minnesota Press, 2010.

Wolfe, Cary. *Zoontologies: The Question of the Animal*. Minneapolis: University of Minnesota Press, 2003.

Zabala, Santiago. *The Remains of Being: Hermeneutic Ontology After Metaphysics*. New York: Columbia University Press, 2009.

Index